Midland Railway and L M S 4-4-0 Locomotives

Their Design, Operation and Performance

Front cover photo:
1119, an LMS 4P Compound built in 1925 and based at Crewe North depot, where it is seen ex-works in May 1937. (Colour Rail)

Front cover inside jacket photo:
LMS 2P 40661 at Kilmarnock, July 1960. It was built in 1931 and was withdrawn in December 1961. (Colour Rail)

Back cover photos:
Midland '1562' class No 1666 (later 357) the last of thirty 1882 Derby built class 2s, as built, c1895. (F.Moore/MLS Collection)

777, one of the remaining 'saturated' class 3 'Belpaires' at the Grouping. 777 was fitted to burn oil during the 1921 miners' strike. It was rebuilt with a superheated boiler in 1924. (E.M.Johnson Collection)

40634, one of the three former LMS 2P 4-4-0s built for the Somerset & Dorset Railway in 1928 and numbered 45, still found on the S&D system at Templecombe in April 1952. (J.Davenport/MLS Collection)

Midland Railway and L M S 4-4-0 Locomotives

Their Design, Operation and Performance

DAVID MAIDMENT

First published in Great Britain in 2021 by
Pen & Sword Transport
An imprint of Pen & Sword Books Ltd
Yorkshire - Philadelphia

Copyright © David Maidment, 2021

ISBN 978 1 52677 250 3

The right of David Maidment to be identified as Author of this work has been asserted by him in accordance with the Copyright, Designs and Patents Act 1988.

A CIP catalogue record for this book is available from the British Library.

All rights reserved. No part of this book may be reproduced or transmitted in any form or by any means, electronic or mechanical including photocopying, recording or by any information storage and retrieval system, without permission from the Publisher in writing.

Typeset in Palatino by SJmagic DESIGN SERVICES, India.
Printed and bound in India by Replika Press Pvt. Ltd.

Pen & Sword Books Ltd incorporates the Imprints of Pen & Sword Books Archaeology, Atlas, Aviation, Battleground, Discovery, Family History, History, Maritime, Military, Naval, Politics, Railways, Select, Transport, True Crime, Fiction, Frontline Books, Leo Cooper, Praetorian Press, Seaforth Publishing, Wharncliffe and White Owl.

For a complete list of Pen & Sword titles please contact:

PEN & SWORD BOOKS LIMITED
47 Church Street, Barnsley, South Yorkshire, S70 2AS, England
E-mail: enquiries@pen-and-sword.co.uk
Website: www.pen-and-sword.co.uk

Or

PEN AND SWORD BOOKS
1950 Lawrence Rd, Havertown, PA 19083, USA
E-mail: Uspen-and-sword@casematepublishers.com
Website: www.penandswordbooks.com

All David Maidment's royalties from this book will be donated to the Railway Children charity [reg. no. 1058991] [www.railwaychildren.org.uk]

Other books by David Maidment:

Novels (Religious historical fiction)
The Child Madonna, Melrose Books, 2009
The Missing Madonna, PublishNation, 2012
The Madonna and her Sons, PublishNation, 2015
The Reluctant Traitor, PublishNation 2021

Novels & Short Stories (Railway fiction)
Lives on the Line, Max Books, 2013
Steamy Stories, PublishNation 2021

Non-fiction (Railways)
The Toss of a Coin, PublishNation, 2014
A Privileged Journey, Pen & Sword, 2015
An Indian Summer of Steam, Pen & Sword, 2015
Great Western Eight-Coupled Heavy Freight Locomotives, Pen & Sword, 2015
Great Western Moguls and Prairies, Pen & Sword, 2016
Southern Urie and Maunsell 2-cylinder 4-6-0s, Pen & Sword, 2016
Great Western Small-Wheeled Double-Framed 4-4-0s, Pen & Sword, 2017
The Development of the German Pacific Locomotive, Pen & Sword, 2017
Great Western Large-Wheeled Double-Framed 4-4-0s, Pen & Sword, 2017
Great Western Counties, 4-4-0s, 4-4-2Ts & 4-6-0s, Pen & Sword, 2018
Southern Maunsell Moguls and Tank Engines, Pen & Sword, 2018
Southern Maunsell 4-4-0s, Pen & Sword, 2019
Great Western Granges, Pen & Sword, 2019
Cambrian Railways Gallery, Pen & Sword, 2019
Great Western Panniers, Pen & Sword, 2019
Great Western Kings, Pen & Sword, 2020
Great Western & Absorbed Railway 0-6-2 Tank Locomotives, Pen & Sword, 2020
Drummond's L&SWR Passenger & Mixed Traffic Locomotives, Pen & Sword, 2020
Southern 0-6-0 Tender Locomotives, Pen & Sword, 2021
LNER 4-6-0 Locomotives, Pen & Sword, 2021

Non-fiction (Street Children)
The Other Railway Children, PublishNation, 2012
Nobody ever listened to me, PublishNation, 2012

CONTENTS

Acknowledgements 7

Introduction 8

Chapter 1   **The Engineers** 12
    S.W. Johnson, 1873-1903 12
    R.M. Deeley, 1904-1909 12
    Sir Henry Fowler 13
        Midland Railway, 1910-1922 13
        LMS, 1923-1931 13

Chapter 2   **The Midland Class 2** 15
    The 1312 class (1907 Renumbering & LMS 300-309) 15
    The 1327 class (310-327) 20
    The 1562 & 1667 classes (328-357) 25
    The 1738 class (358-377) 38
    The 1808 class (378-402) 45
    The 2183 class (403-427) 57
    The 2203 class (428-472) 66
    The 2581 class (473-482) 79
    The 1667 replacement class (483–492) 83
    The 150 class (493-502) 89
    The 2421 class (503-522) 93
    The 60 class (523-562) 97
    The Somerset & Dorset 4-4-0s 109
    The Midland & Great Northern 4-4-0s 120

Chapter 3   **The Midland Class 3** 124

Chapter 4   **The Midland Class 4** 158
    The Smith/Deeley Compound 158
    The '990' class 194

Chapter 5	The LMS Options	204
Chapter 6	The LMS 4P Compound	211
Chapter 7	The LMS 2P	265
Chapter 8	Conclusions	299
Appendix	Dimensions, Weight Diagrams & Statistics	302
	Midland Main line Gradient Charts	344
	Bibliography	347
	Index	348

ACKNOWLEDGEMENTS

My membership of the Stockport station based Manchester Locomotive Society has been particularly valuable in both the research for text and photographs for this book. Their extensive library has a large number of books about Midland Railway and LMS locomotives, details of which are listed in the bibliography. The Summerson volumes about Midland locomotives give far more technical detail than I am qualified to expound for those whose main interest this is, while I concentrate more on the overall history and performance, as well as showing a large number of photographs of the many variations which I am hopeful that modellers will find of value. The library contains full sets of magazines such as the *Railway Magazine* and *Trains Illustrated* which I have scoured at leisure and I'm indebted also to the Railway Performance Society who have a very large collection of performance logs easily accessible. Readers will soon notice that nearly all of the photographs in the book are from the Manchester Locomotive Society (MLS) archive and I thank the Society and their photo archivist, Paul Shackcloth, for access to them and permission to publish them free of any fee as the book royalties are being donated to the Railway Children charity. I founded this charity with the help of many within the railway industry in 1995 to protect and support very vulnerable street children found on the transport systems of India, East Africa and, unfortunately, even – far too often – runaway children on the railway stations of our own country (www.railwaychildren.org.uk.) I also thank E.M (Eddie) Johnson, a member of MLS, who has provided me with a number of excellent photographs from his collection on the same basis.. Many of the photographs I've used from the MLS collection unfortunately lack details on the back of the print so that I'm not always able to acknowledge the original photographer in the credit to the caption. If there are copyright holders I've been unable to trace, please contact the publisher.

I'm also grateful to those in the Pen & Sword team who have been most supportive as ever – John Scott-Morgan, Commissioning Editor and friend, Janet Brookes, Production Manager, Carol Trow, my editor, and the design and marketing team with whom it's been a pleasure to work.

INTRODUCTION

I've been persuaded once more to write a book about locomotives from other than the Great Western or Southern railways. I have hitherto insisted on writing about locomotives I knew and travelled behind in leisure or work and could therefore incorporate some of my own personal experiences. I did indeed work on the London Midland Region of British Rail, but not until the 1980s and then only in the lofty position of Chief Operating Manager, which did at least provide me the opportunity to savour a few footplate runs allegedly in the course of my duties to check on the safety of the steam specials on the Settle & Carlisle where I was guest on the footplate of 46229 *Duchess of Hamilton* and on Southern and Eastern pacifics on the Banbury-Marylebone route. I also remembered my responsibility for the service run by BR's last three steam engines in operating stock – the Vale of Rheidol's narrow-gauge GW 2-6-2 tanks on which I rode No 7 *Owain Glyndwr*.

My experience of the Midland and LMS 4-4-0s is, however, regrettably sparse. My first encounter was a distinct surprise. I was with my family on our fortnight's summer holiday in Bournemouth and took a few hours off from the beach to trainspot at Bournemouth Central. I paid just one visit to Bournemouth West which I found somewhat underwhelming until the apparition of 40601 backing onto the three Southern corridor coaches standing in the otherwise empty station. My knowledge of the Somerset & Dorset Railway advanced from nil to 1 per cent and I took a photograph of it with my simple Kodak camera, pictured here.

My introduction to its larger companion, the LMS Compound, took place during train spotting trips to the capital – at St Pancras in 1950 when I encountered 41049 that had piloted a 'Black 5' on an incoming express and in 1951 at the end of platforms 12 and 13 at Euston, when between two 'Royal Scots' protruding from the ends of the platforms at the head of the summer Saturday *Welshman* and its relief, I spied a very clean 41144 arriving on the far side. My fellow spotters explained to the ignorant twelve-year-old what this was and the fact that it was said by them to be shedded at Nottingham. I've since confirmed they were right, though what it was doing at Euston I have no idea.

LMS 2P 40601 based at Bath Green Park on an afternoon stopping train to Bath, 13 May 1951. This was (although I certainly didn't know it at the time) the 2P that received double port exhaust valves. (David Maidment)

On later London spotting trips, I ventured out to suburban stations like Willesden Junction and Hendon where I saw both class 2Ps and LMS Compounds piloting Black 5s and Jubilees on expresses. My sympathies lay with the crews of these 4-4-0s which seemed to be swinging wildly as the expresses sped London-wards. I remember in particular an LMS 2P in front of a Jubilee tearing through Willesden Junction appearing to lurch to the right as the train and train engine took the left hand curve at an estimated 80mph+.

In the later 1950s, my experiences were more mundane. After a week on a Western Region 'short works experience' course at Bristol when I was in the Sixth Form, I spent a couple of days visiting an elderly great aunt in Wickwar on the Midland Bristol-Gloucester line, and found my four-coach stopping train standing in the old Brunel part of Temple Meads steaming gently behind Bourneville's 41073 which made a great fuss of hauling its light load up the 1 in 69 of Fishponds bank, steam oozing from all orifices. Visits with the Charterhouse Railway Society to Kentish Town in 1956 produced a couple of Compounds dead on shed, and I remember a number of us clambering over the footplate of rusting 41103, although back at St Pancras a smarter 41199 was blowing off steam at the head of a semi-fast late afternoon train to Bedford.

Then in 1957, between school and college, I worked at Old Oak Common, travelling daily from my home in Woking. The quickest way from Waterloo to Willesden Junction, the nearest station to Old Oak, was via the Bakerloo line direct, but I soon discovered that my finish time coincided with a Rugby-Euston semi-fast train, connecting there with the Northern Line. During the winter and spring of 1957 the train was invariably hauled by one of Rugby's Black 5s, but from the beginning of the summer timetable, they were utilised on higher priority turns and several of Rugby's Compounds were taken out of store and one was diagrammed to my train home. I soon discovered that I had to be very patient to enjoy this privilege, for the train was rarely on time, anything from 10 to 25 minutes late. No 41093, recently transferred from Llandudno Junction, was one of the better ones and turned up four times. 41162 was a regular, and 41105, 41113, 41122 and 41172 turned up from time to time.

Occasionally, the Compound was replaced by a Camden Jubilee and I soon established that *Fisher* or *Anson* had been 1B's standby engine the previous night covering for the Rugby Compound's return trip failure. I can vividly remember the one night 41122 surprised everyone by appearing on time and actually ran into platform 7 at Euston half a minute early. The platform was festooned with flags which seemed a little over the top, until I found out that a royal train had departed from that platform half an hour earlier! The last home run I had that summer was a role reversal. An express limped into Willesden Junction's up slow platform with a Longsight Compound, 41168, coming to the rescue of ailing Jubilee, 45678, *De Robeck*, which I boarded hurriedly for the five mile run to Euston.

Rugby Compound 41113 pauses at Willesden Junction on the 6.09pm to Euston semi-fast from Rugby, where it will stand for ten minutes to allow ticket inspection as the train will run into an open platform at Euston, 1 August 1957. (David Maidment)

Compound 41162 heads a semi-fast train from Rugby to Euston via Northampton, passing 45187 on a Rugby-Euston stopping passenger train routed via Weedon and the direct line, c1957. (Author's Collection)

By the following summer they'd gone, and my only other experience was in 1980 with the restored 1000 double-heading the preserved V2, 4771 *Green Arrow*, on a railtour between Hellifield and Carnforth (see page 194). Beautiful though this engine is, albeit now just a static exhibit, I have to admit to a grudging affection for the run-down Compounds. In those days, I happily accepted the uncertainty of the long wait at Willesden Junction in the hope of a Compound rather than 44712 or 44716 yet again (but on time)!

I did see some of the old Midland engines. Back in 1950, on a three week vacation with a Yorkshire school friend at Doncaster, we made bus trips to trainspot in various South Yorkshire locations. One outing took us to Sheffield and after a wasted visit to Sheffield Victoria where we saw one B1 (not even a 'cop'), we hastened to the Midland station, where we noted two of the Midland Compounds, 41015 and 41019, though their comparative rarity at that time was lost on us. Many years later, passing through Derby, I took a photo of one of the much-rebuilt Johnson 2P 4-4-0s, 40412 (Midland 2183 class of 1892).

I did not discover then the complex history of 40412 and its sisters and it is only now

Midland rebuilt Johnson '2' 40412 of the '2183' class, alongside 4F 0-6-0 44420, at Derby station, 7 September 1957. It was withdrawn from Derby shed in May 1959. (David Maidment)

that I am trying to unravel their development and differences – wheel diameters, boiler renewals, superheating – and try to help those of you that are modellers as well as enthusiasts to identify the right details and liveries for the various periods of their existence. In addition to the 2Ps and 4P Compounds and their LMS successors, I also describe the history and operation of the less well known 3P Midland 700 class and the ten 'simple' 4P 4-4-0s (990 class) which spent most of their operating activity on the Settle & Carlisle route. In addition to the history of the designs, construction and rebuilding, I have researched the archives of the Rail Performance Society and back numbers of the *Railway Magazine* and *Trains Illustrated* to give an idea of their performance in the halcyon days of the Midland Railway, the early 'small engine' days of the LMS, and a few remnants of their main line work in the 1940s and '50s as well as an account of 1000 in the preservation era. Two other railways built or acquired Johnson or Fowler class 2 4-4-0s or locomotives developed from them – the Somerset & Dorset Railway and the Midland & Great Northern Railway and I also cover a brief history of these engines and their subsequent acquisition by the LMS and LNER.

Chapter 1
THE ENGINEERS

S.W. Johnson, 1873-1903

Samuel Waite Johnson was born in Bramley, Leeds, on 14 October 1831, the son of James Johnson, an engineer who later worked for the Great Northern Railway. Samuel was educated at Leeds Grammar School and became an engineering pupil of James Fenton at the Leeds firm of E.B. Wilson where, amongst other things, he assisted David Joy in producing the drawings of the famous *Jenny Lind* steam locomotive. He was also involved in the construction of the 'Bloomer' 0-6-0 locomotives for the London & North Western Railway. His first appointment was as Assistant District Locomotive Superintendent of the Southern Area of the GNR, then Works Manager at Peterborough and in 1859 he became Acting Locomotive Superintendent of the Manchester, Sheffield and Lincolnshire Railway, one of the companies that later formed the Great Central. In 1864, he moved to Cowlairs, Glasgow, as Locomotive Superintendent of the Edinburgh & Glasgow Railway which was amalgamated with the North British Railway in 1865. For his service of training pupils for the Egyptian railways at Cowlairs, the Viceroy of Egypt made him a Commander of the Order of Medjidie and an Officer of the Order of Osmanie. In 1866, he went south to Stratford as Locomotive Superintendent of the Great Eastern Railway after the resignation for health reasons of Robert Sinclair. The GER Board Meeting of 16 July 1873 recorded:

> Resolved: that Mr Samuel Waite Johnson having been engaged as Locomotive Superintendent of the Great Eastern Railway for a period of upwards of seven years, the Directors desire to testify their complete satisfaction with his Engineering abilities and with the manner in which he has discharged the duties of his office.

With this glowing testimonial, he was chosen from twenty-six applicants by the Directors of the Midland Railway to be appointed as Locomotive Superintendent after the death of Matthew Kirtley, at an initial salary of £2,000 a year (compared with £750 on the GER) raised to £2,500 a couple of years later in 1875. He remained in charge of locomotive matters for over thirty years.

His initial designs for the Midland were 2-4-0s based on Kirtley's engines but with inside frames. He designed and built numerous classes of 0-6-0s, 0-4-4 tank engines and the famous Midland 'Singles', 4-2-2s, one of which is displayed at the York National Railway Museum. He produced his first 4-4-0 design in 1876 and numerous developments of this throughout his period of office, which are the subject of this book.

Not a lot is known about the character of the man. He is said to have been reserved, though he must have been strong enough to command a Works like Derby. He combined his engineering knowledge with an eye for artistry as his locomotive lines and liveries generated great admiration. Samuel Johnson retired at the end of 1903, aged 72, and lived at Nottingham out of the limelight where he was a Justice of the Peace and was very involved with St Peter's Church there. He became a Member of the Institute of Mechanical Engineers in 1861, the Institute of Civil Engineers in 1867, a Council Member of the IMechE in 1884, Vice President in 1895 and President in 1898. His son, James, followed in his father's footsteps, becoming Locomotive Superintendent of the Great North of Scotland Railway in 1890. Samuel died in 1912, aged 80.

R.M. Deeley, 1904-1909

Richard Mountford Deeley was born in Chester on 24 October 1855 and was educated at the Chester Cathedral Grammar School.

In 1873, he became an engineering pupil of Edward Ellington, the Managing Director of the Hydraulic Engineering Company, Chester, and was sent in 1874 to work on hydraulic engines with Brotherhood & Hardingham, a London company. He left to become a pupil of Samuel Johnson at Derby in 1875 and on the completion of his apprenticeship there in 1880 was appointed Chief of Testing. In 1893, he was appointed as Inspector of Locomotives and in 1899 supervised the design and maintenance of all electrical plant on the Midland Railway. He was appointed Derby Locomotive Works Manager in 1902 and became Locomotive Superintendent at the beginning of 1904 on Samuel Johnson's retirement. The Midland Board decided on a reorganisation of the Locomotive Department in 1909, creating a separate Locomotive Running Department under the Traffic Department. Some internal politicking is hinted at, and Deeley consequently resigned, although he was clearly highly regarded by the Directors, who minuted at their Board Meeting of August 1909:

> 'Owing to certain contemplated changes in the Locomotive Department, Mr Deeley had placed his resignation in his (the Chairman's) hands. It was resolved that Mr Deeley's resignation be accepted from 30 November 1909 and that he be granted a retirement allowance of £1,200 per annum during the pleasure of the Directors. Also that Mr Deeley be requested to retain his gold pass over the Company system for life.'

He became a Member of the Institute of Mechanical Engineers in 1890 and the Institute of Civil Engineers in 1906. After his resignation, he wrote books on aspects of engineering, meteorology and genealogy. He died in Isleworth, London, on 19 June 1944, aged 88.

Sir Henry Fowler
Midland Railway, 1910-1922

Henry Fowler was born at Evesham in Worcestershire on 29 July 1870 and was educated at Evesham Grammar School. Between 1885 and 1887, he attended the Mason Science College in Birmingham and for the following four years was apprenticed under John Aspinall at Horwich Works of the Lancashire & Yorkshire Railway Company. He was awarded the first Whitworth Exhibition at the Horwich Mechanics Institute in 1891 and became a teacher there. He became at the same time Assistant to George Hughes in the Locomotive Testing Department, succeeding him when the latter was promoted. In 1895, he was appointed to be the L&Y's Gas Engineer. He was also engaged in automobile engineering (much later in 1920 he was elected President of the Institute of Automobile Engineers).

He moved from Horwich in 1900 to become Gas Engineer of the Midland Railway at Derby, and in 1905 was appointed Assistant Works Manager there, becoming Works Manager in 1907. When Deeley resigned in 1909, Fowler was appointed as Chief Mechanical Engineer in charge of locomotive design, construction and maintenance but now stripped of the locomotive running management which had been passed to the Traffic Department. Fowler was basically an all-round Engineering Manager rather than a locomotive designer, the detailed drawings and development carried on very much by the design staff in the Drawing Office at Derby.

In 1915, he was appointed as Director of Production to the Ministry of Munitions and in 1917, Assistant Director General of aircraft production. In consequence of his wartime service, he was awarded the CBE in 1917 and was knighted in 1918.

LMS, 1923-1931

He became deputy Chief Mechanical Engineer to the LMS in 1923 under George Hughes, who retired after two years. Fowler therefore succeeded him in 1925 and moved the LMS motive power headquarters from Horwich to Derby, continuing the 'small engine' policy of the former Midland's Traffic supremo, Anderson, no doubt to the indignation and resentment of the Crewe based ex-LNWR staff. Fowler's strengths were in engineering production management and he reorganised Derby Works with many economic benefits. Locomotive design was left to Herbert Chambers, Chief Draughtsman at Derby, and Anderson who was Motive Power Superintendent overseeing all locomotive running and performance matters.

However, the LMS needed more powerful express locomotives, especially for the heavy loaded trains on the West Coast main line, resulting in the LMS Board testing a GWR 'Castle', then requesting fifty of the class, which regrettably the GWR declined. Fowler then obtained drawings of the Southern Railway's new 'Lord Nelson' class and contracted the North British Company to design a similar powered engine, though of three rather than four cylinders. The resultant 'Royal Scot' class was delivered in 1927, though the lessons from the GWR 'Castle' exchange had not been fully learned as the new engines suffered from the Midland Railway small bearing design and outdated valve gear – design flaws that marred Fowler's Beyer Garretts for the heavy Midland coal traffic and which Stanier sorted with the rebuilding of the Royal Scots in the 1930s. The first 'Royal Scot' to be rebuilt, 6103, emerged from Crewe Works in 1943. The successful 2-6-4 tank was designed under his leadership, but after the failure of the high-pressure experimental engine Fury in 1930, Fowler was moved sideways to become Assistant to the Vice-President of Research & Development, Sir Harold Hartley, with Ernest Lemon covering the CME post until the appointment of William Stanier from the GWR in 1932.

Fowler was more in his element researching metallurgy and in 1932 was elected as President of the Institute of Metals. He was also skilled in the training of young engineers and was a member of the governing body of the Midland Institute in Derby. He was awarded an honorary LLD by Birmingham University and a DSc by Manchester University and was a Member of its Council from 1928 to 1934.

Henry Fowler had married Emma Needham in 1895 and the pair had two sons and a daughter. She died in 1934 and Sir Henry died on 16 October 1938 at Spondon Hall, Derby.

Chapter 2
THE MIDLAND CLASS 2

The 1312 class (Midland 1907 & LMS 300-309)
Design & construction

The Midland Railway Locomotive Committee was notified of a requirement for 'ten express bogie engines for the Settle and Carlisle line' on 18 January 1876 and Johnson, who had designed and built two inside cylinder 4-4-0s for the Great Eastern Railway, produced a similar design to satisfy the need, putting them out to tender to six companies. Kitson's tender at £2,750 per locomotive was accepted in February and the ten engines, numbered 1312-1321, were delivered between November 1876 and February 1877, and were identified as class 'F'.

The key dimensions of the ten locomotives were:

Cylinders (2 inside)	17½ x 26in
Coupled wheel diameter	6ft 6in
Bogie wheel diameter	3ft 3in
Boiler pressure	140lb psi
Heating surface	1,215sq ft
Grate area	17.5sq ft
Axleweight	14 tons
Weight: Engine	41 tons 19 cwt
Tender	35 tons 3 cwt
Total	77 tons 2 cwt
Water capacity	2,800 gallons
Coal capacity	3½ tons
Tractive effort	10,800lbs

The slim-line boiler was classified the 'B' type. They only had tender handbrakes initially but were vacuum fitted in 1877. They appeared in Kirtley's dark Brunswick green livery, though from 1877 Johnson introduced a lighter green. Then, in October 1883, Johnson tested a red oxide paint and when proven successful, with a potential saving of £2,000 per annum for the fleet, this was adopted and was the well-known Midland crimson red livery. The class was renumbered 300-309 at the Midland Railway renumbering scheme of 1907.

Although many of the Midland 4-4-0 classes were later rebuilt with larger boilers, the 1312 class remained unchanged apart from enlargement of the cylinders to 18in diameter between 1884 and 1891.

Two were withdrawn between 1910 and 1912 -307 and 309 – but the rest survived because of increased traffic demands in wartime.

No 1314 in Works grey as built in 1876. It was withdrawn in January 1920. (MLS Collection)

No 1320 at Derby in the original Midland Johnson light green livery, c1882. (E.J. Johnson Collection)

The prototype Midland Railway 4-4-0, 1312, in the Johnson oxide red livery at Liverpool Brunswick shed, c1890. It was withdrawn by the LMS in December 1924. (MLS Collection)

No 300, the former 1312 renumbered in 1907, with Midland Railway crest on the cab and numerals on the tender and smokebox door, at Leeds, c1910.
(MLS Collection)

No 306, the former 1318, taking water on shed, c1912. This was the last survivor of the class, withdrawn by the LMS in November 1930.
(Real Photographs/MLS Collection)

No 301 (ex 1313) on shed shortly before withdrawal after the First World War, c1919. Note military personnel on the footplate. (MLS Collection)

Nos 301 and 302 were withdrawn at the end of the war, the six remaining being taken over by the LMS in 1923. No 306 (the former 1318) was not withdrawn until November 1930.

Operations

The Settle and Carlisle line was opened on 1 May 1876 and as the new 4-4-0s were not ready, the 2-4-0s ordered in 1875 were utilised. After their construction, the 4-4-0s did not work the line for which they were intended and by 1880 all ten were allocated to the Cornbrook depot at Manchester. By 1892, 1312-1317 were based at the Lancashire & Yorkshire depot at Sandhills working from Liverpool and Manchester over L&Y rails to Hellifield connecting there with the Settle and Carlisle line. Nos 1318-1321 were based at Liverpool Brunswick working Liverpool-Blackburn portions of Midland expresses and services on the Cheshire Lines Committee routes. All remaining engines were based at Liverpool during the First World War before being transferred to Sheffield for piloting heavy coal trains between Masborough and Toton. The last survivor was transferred to Derby in 1928 replacing the Johnson 'single' 600, on inspection and VIP saloon duties.

No 306 (ex 1318) on a passenger train at Hough Green, c1912. (MLS Collection)

No 308 (ex 1320) still at work on a local passenger train, c1920. (E.M. Johnson Collection)

The 1327 class (310-327)

Design & construction

Further motive power needs were identified by Samuel Johnson in July 1876, the ten already ordered being deemed insufficient. The construction and supply of a further twenty 4-4-0s was therefore put out to tender and Dübs was selected out of fourteen bids, the cost per locomotive being quoted at £2,495. The twenty engines, delivered between June and November 1877, were numbered 1327-1346 and classified 'G'. They were similar in many ways to the 1312 class 'F', the main dimension differences being:

Cylinders (2 inside)	18in x 26in
Coupled wheel diameter	7ft 0in
Bogie wheel diameter	3ft 6in
Heating surface	1,313sq ft (later changed to 1,260sq ft)
Water capacity	2,960 gallons

Boiler type, brake arrangements and livery were similar to those of the 1312 'F' class. Again, like the 'F's, they retained the type 'B' boiler to the end. Two of the class were withdrawn as early as 1904 before renumbering, 1332 and 1336, and 1346, the last of the class, took the 1336 number space in August 1904. Four, 311, 319, 323 and 326, received new frames and three of these (excluding 326) had 160lb psi type 'B' boilers at the later stages of their lives.

The eighteen remaining engines were renumbered 310-327 in 1907. Nos 315, 316 and 324 were withdrawn before the First World War and another five, 312, 313, 318, 321 and 322, between 1918 and 1920. Ten were taken into LMS ownership and 311 (formerly 1328) outlasted the rest by several years, being finally condemned in November 1934.

The rebuilt 323 (ex 1341) with new frames, higher pressure boiler and extended smokebox, at King's Norton on a Birmingham-Gloucester stopping service, 21 August 1912. (W.L. Good/MLS Collection)

No 1332 at Kentish Town, c1880. This engine was one of the two withdrawn in 1904 before renumbering. (MLS Collection)

The rebuilt 323 in the LMS era when it was stationed at Toton for freight work, seen here at its home depot, c1927. It was withdrawn in December 1928. (E.M. Johnson Collection)

Operations

The first fourteen of the new continuously-braked locomotives were found on the Manchester-London expresses, either between Manchester and Leicester or Leicester and St Pancras. The remaining six went to Leeds. A couple of years later, twelve of the early ones were based in London and 1341-1345 were still at Leeds but 1339 was at Derby and 1346 at Liverpool. As newer 4-4-0s were built, the London engines migrated to Manchester, as did the Leeds engines, with Derby's 1339 joining 1346 at Liverpool.

By 1892 the distribution of the twenty locomotives was as follows:

Manchester Cornbrook:	1327-1335, 1340, 1341
Birmingham Saltley:	1336-1339
Liverpool Brunswick:	1342, 1343
Leeds:	1344
Skipton:	1345, 1346

The first two withdrawals took place in 1904 (1332 and 1336) and the succession of 4-4-0 classes designed and constructed by Johnson had removed the 1327 class from the main express work. In 1908, their allocation (using the revised 1907 numbers) was as follows:

Manchester Cornbrook:	315-317 (1333-1335)
Skipton:	310-314 (1327-1331)
Sheffield Millhouses:	318-327 (1337-1346*)

(*1346 had been renumbered 1336 after that locomotive's withdrawal)

One of the class that received new frames and a 160lb psi boiler in 1909 went to Derby and in 1917 to Toton where it was used, with 301 of the 1312 class, to act as pilot to heavy coal trains to Wellingborough and Cricklewood.

An early photograph of 1878-built 1327 class No 1335 on a passenger train crossing the River Lune on the Midland Railway 'Greyhound Bridge' at Lancaster with a train from Leeds and Bradford for Morecambe, c1890. (E.M. Johnson Collection)

Sheffield's 1338 on an up express near Chinley, c1904. (MLS Collection)

Sheffield's 325 (ex 1343) on an up express leaving Leicester, c1912. (A.C. Gilbert/ MLS Collection)

Sheffield's 322 (ex 1340) piloting a Midland 3F 0-6-0 on a down mineral train leaving Shotlock Tunnel on the final climb to Ais Gill, 1918. Note one of the crew attending to the motion of the 4-4-0, presumably with an oil can. (E.S. Cox/MLS Collection)

No 327 (ex 1345) leaving Derby with a northbound freight, passing a Midland 0-4-4T, 13 August 1920. (H.C. Casserley/MLS Collection)

The last survivor, 311 (ex 1328), enters Rugby station with the 1.55pm from Peterborough East, whilst an ex-LNWR express locomotive hauled train waits to depart for Euston, 12 May 1934. (W. Potter/MLS Collection)

After the Grouping, 320 spent a couple of years at Burton and the last survivor, 311, was at Skipton until 1928, when it moved briefly to Liverpool before going to Nottingham in 1929. It was finally used from Peterborough shed in 1933, working the branches to Rugby and Northampton, before final withdrawal in November 1934, having outlived its other sisters by over six years.

The 1562 & 1667 classes (328-357)
Design & construction

Johnson had reverted to a 2-4-0 design in 1880/1, but by 1882 had decided that future construction for express work should be the four-coupled bogie design. The Midland Railway train mileage was increasing, and the Board authorised ten new 4-4-0s to be constructed at Derby at an estimated cost of £2,200 each, significantly cheaper than those constructed by contractors. However, actual costs were calculated later as in excess of £2,800. Nos 1562-1571 were built between September and November 1882. A September 1892 report on motive power requirements identified a shortfall of thirty-five main line engines, so a further thirty 4-4-0s were authorised (and five tank engines). They were built at Derby at an authorised cost of £2,300 each (estimated later as an actual cost exceeding £2,600), with 1572-1581 being delivered between February and April 1883, 1657-1666 between October and November of the same year. A further ten, 1667-1676, were constructed between May and October 1884. The key dimensions of the thirty 1562 class locomotives were:

Cylinders (2 inside)	18in x 26in
Coupled wheel diameter	6ft 9in
Bogie wheel diameter	3ft 6in
Stephenson valve gear	
Boiler pressure	'B' type 140lb psi
Heating surface	1,260sq ft
Grate area	17.5sq ft
Axleweight	15 tons 14 cwt

Weight:	Engine	41 tons 19 cwt
	Tender	35 tons 3 cwt
	Total	77 tons 2 cwt
Water capacity		2,950 gallons
Coal capacity		3½ tons
Tractive effort		14,560lbs

1572-1581 were initially fitted with Westinghouse automatic air brakes, the remainder with vacuum brakes, though the Westinghouse equipment was replaced in the 1890s. Nos 1675 and 1676 were also fitted with the Westinghouse air brake system.

The ten 1667 class engines had 7ft 0½ in coupled wheels, 19in diameter cylinders and were fitted with Joy's valve gear and overhead slide valves. They were also equipped with boilers pressured at 160lbs psi. Nos 1670 and 1672 received piston valves in 1890. No 1672 was withdrawn in 1896 and 1670 reverted to slide valves in 1897. The attempt to provide a more powerful engine was unsuccessful and they were withdrawn between 1896 and 1901 and replaced by new engines given the same numbers, renumbered 483-492 in 1907.

Johnson developed a larger boiler for his later 6ft 6in 4-4-0s in

No 1562 as built new in 1882. (Real Photographs/MLS Collection)

No 1666, the last of the thirty 1882 Derby built class 2s, as built, seen in the Midland Railway red livery, c1895. (F. Moore/MLS Collection)

No 1579 as built new in 1883, seen here c1890. (E.M. Johnson Collection)

No 1670 as built in 1884 with 160lb psi boiler and Joy's valve gear, c1890. No 1670 was withdrawn in April 1901. It was replaced by a locomotive bearing the same number. (MLS Collection)

No 1673 as built in 1884 with 160lb psi boiler and Joy's valve gear, c1890. No 1673 was withdrawn in May 1901. (F. Moore/MLS Collection)

No 1668 at Manningham, Bradford, a year before withdrawal and replacing with a conventional Johnson 4-4-0 with the identical number, c1895. (E.M. Johnson Collection)

1903, classified as 'H'. Although a decision was made to fit a significant number of the 4-4-0s with the H boiler in Johnson's last year, the rebuilding took place during Deeley's time in charge at Derby. All of the earlier class 1562 engines and the replacement 1667 class were subsequently fitted with this boiler between 1906 and 1908. These boilers were pressed at 175lbs psi, heating surface of 1,428sq ft and grate area of 21.1sq ft. Some had a later version classified 'HX' which had a slightly smaller heating surface of 1,347sq ft. Many had renewed frame provision at the same time as the reboilering.

However, Deeley had developed the G7 boiler intended originally for a new series of 0-6-0s, a version of the 'HX' boiler with Belpaire firebox. It was also pressed at 175lbs psi and had a total heating surface of 1,283sq ft. A decision was taken in December 1908 to replace the 'H' boiler on all the classes by the G7 and sixteen 1562 class locomotives, 1562-1564, 1568, 1570-1574, 1576, 1580, 1581, 1558-1660, 1665, received replacement boilers between 1909 and 1911. The thirty 1562 class engines had been renumbered 328-357. Some of these engines carried the H boiler before further rebuilding for a very short time – 336 for only twenty-one months.

Finally, a decision was made to rebuild many of the earlier class 2 4-4-0s in similar fashion to the '483' superheated boiler class. Just five of the 1562 class were converted in

No 335, the former 1569, after rebuilding with an 'H' larger boiler in 1907, at Derby in early LMS livery, c1924. No 335 was withdrawn in 1927. (MLS Collection)

No 331 (ex 1565), rebuilt with an 'H' boiler in 1907, was the sole example of the class to acquire an extended smokebox with an 'HX' boiler seen here immediately after its fitting at Derby, December 1925. It was withdrawn in 1928. (MLS Collection)

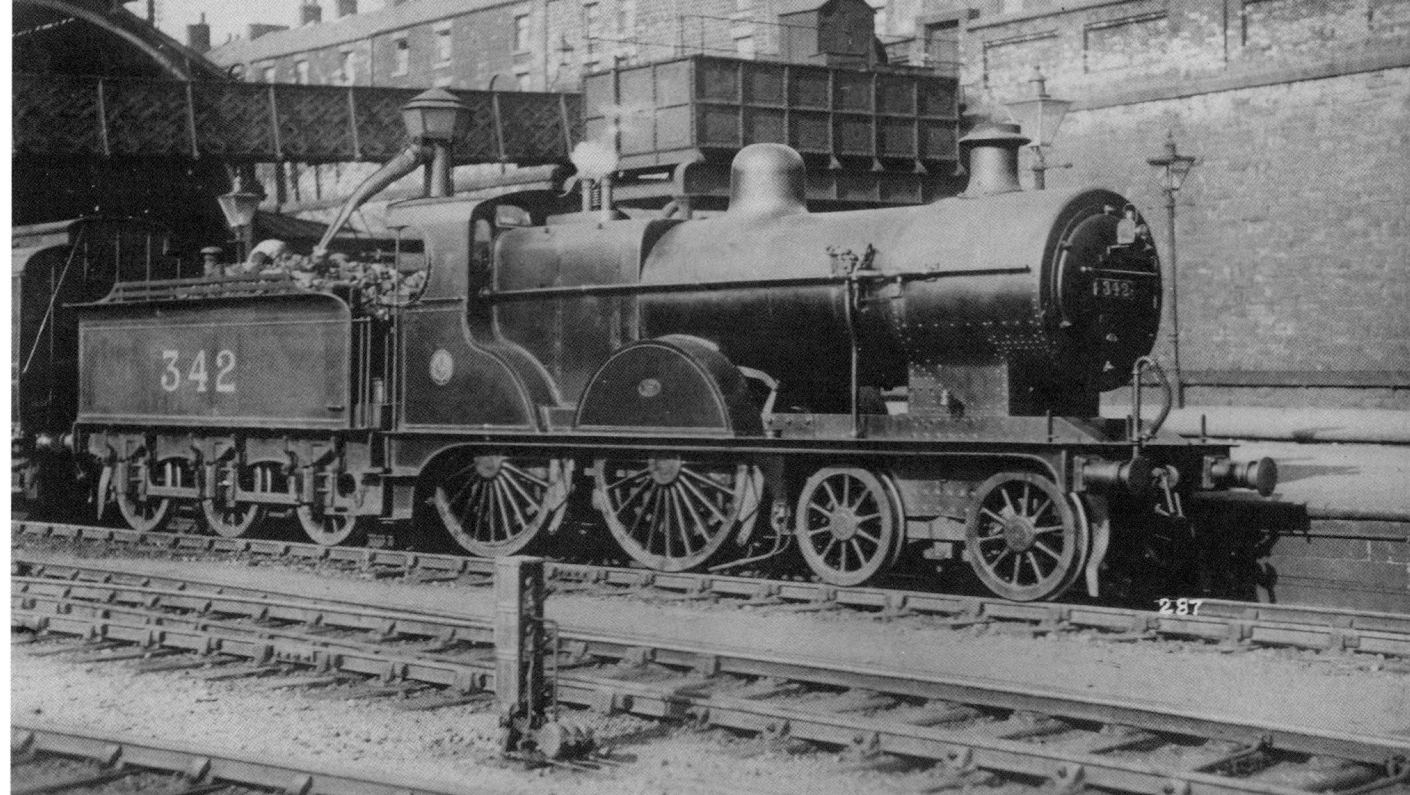

No 342 (ex 1576), rebuilt with an 'H' boiler in 1907 and a G7 boiler/Belpaire firebox in 1911, at Sheffield Midland on a local stopping passenger train, c1932. No 342 was withdrawn the following year. (Railway Photographs/MLS Collection)

No 40332, one of the five '1562' class, formerly 1566, rebuilt with an 'H' boiler in November 1906 and a superheated Belpaire '483' class boiler in 1923, on station pilot duty at Birmingham New Street, c1957. It was the last survivor of the class, withdrawn at the end of 1959. (MLS Collection)

1923 – 332 (ex 1566), 337 (1571), 351 (1660), 353 (1662) and 356 (1665).

The 1562 class with G7 boilers were mostly withdrawn between 1926 and 1930 with 330, 334, 336, 347 and 350 in 1932, 342 in 1933 and the last, 338 (1572), in February 1937. The five rebuilt to '483' class with superheated boilers were not withdrawn until 1953 (40351 & 40353), 1957 (40356), 1958 (40337) and 40332 survived until December 1959. The 1667 class replacement engines were rebuilt as the superheated '483' class in 1912 and 1913, most lasting until the 1950s with 40487 and 40491 lasting into the 1960s (see later, pages 86-88).

Operations

The forty new 4-4-0s quickly replaced the 1876-8 engines on the main expresses and were allocated to the main Midland sheds as follows:

Leicester:	1562 – 1567
Kentish Town:	1568 – 1571
Carlisle:	1572 – 1581
Manchester:	1657 – 1666

The ten 1676 class with 7ft driving wheels were allocated to Kentish Town (5), Nottingham (3) and Derby (2). Thus, all the main line Midland services were covered by these forty new locomotives except for the Derby-Birmingham-Bristol route.

As more 4-4-0s were delivered in the late 1880s and early 1890s, the older engines were cascaded to other services and routes, Leicester and Kentish Town losing their allocation, with four going to Saltley and six to Derby, though Carlisle and Manchester had retained their engines at this period. The ten Joy's valve gear 1667 class had proven a disappointment – they could apparently run fast but had struggled to maintain time with the normal loading of the best expresses. London lost all but two

One of the Kentish Town 1667 Joy's valve 7ft 4-4-0s departing St Pancras with a northbound express for Leicester, c1890. (MLS Collection)

(1675 and 1676) with 1667-1671 being transferred to Manchester. By 1892, further delivery of 4-4-0s dispersed the earlier engines further and the first ten, 1562-1571, were all at Derby. Sheffield, Skipton and Hellifield had acquired engines from the former Carlisle allocation, but Manchester had retained its original ten.

The first *Railway Magazine* volumes were not published until 1897 and it was several months later before Charles Rous-Marten began his monthly series of *British Locomotives' Practice and Performance* in which logs of locomotive running could be found. By that time, it was the later 4-4-0s and class 3s that appeared in these pages, although one article at the end of the magazine's first decade did record the running of one of Manchester's rebuilt 1562 class, 349.

Derby – Manchester, c1908
10.25am St Pancras – Manchester
349 (ex 1583)
187/200 tons

Miles	Location	Times	Speeds		Gradients
0	Derby	00.00			
5.3	Duffield	09.40	pws		
10.4	Ambergate	16.45	sigs	1 ¾ L	
17.2	Matlock	25.15			1/177 R
21.6	Rowsley	30.05		1 L	
25	Bakewell	35.10			1/102 R
27.3	Longstone	39.25			1/100 R
31.4	Millers Dale	46.30	32	2½ L	1/100 R
36	Peak Forest	55.35	28	2½ L	1/90 R
41.6	Chinley	62.15		3¼ L	1/90 F
44.3	New Mills S. Jcn	64.45		2¾ L	1/89 F
53.4	Cheadle Heath	72.50	75	1¾ L	1/100 F
59.2	Chorlton	77.55	sig	T	
61.4	Manchester	84.40	(82 net)	¼ E	

R.E. Charlewood, in a further article in the 1908 *Railway Magazine*, quoted a number of runs on the Midland Railway. Two were with 4-4-0s of this class, both with 1657, which had been rebuilt with the larger 'H' boiler in December 1906, when either the class 3s or Compounds would normally have been rostered to the 1907 105 minute schedule for the 99 miles between Leicester and St Pancras.

Leicester – St Pancras, 1907

1657 — 195 tons | 1657 — 195 tons

Miles	Location	Times	Speeds	Times	Speeds	Gradients
0	Leicester	00.00		00.00		
3.7	Wigston	06.17		06.21		
7.3	Great Glen	10.53	50	11.20	45	1/199 R
10.2	Kibworth	13.58		14.34		
16	Market Harborough	19.29	69	20.26	65	1/238 F
20.8	Desborough	25.51	41/pws 30*	27.03	36/pws 30*	1/132 R
27	Kettering	32.20	70	33.53		1/118 F
34	Wellingborough	38.24		39.46	71/60*	
36.3	Irchester	40.41		41.49		

		Leicester – St Pancras, 1907				
		1657		1657		
		195 tons		195 tons		
Miles	Location	Times	Speeds	Times	Speeds	Gradients
	Sharnbrook summit	-	43	-	40	1/120 R
42.3	Sharnbrook	47.30	68	49.24		1/119 F
49.2	Bedford	53.57	60*	55.17	76½	
57.3	Ampthill	63.09	47	63.41	51	1/200 R
61.8	Harlington	68.29		68.46		1/200 R
66.3	Leagrave	74.08	45	74.17	47½	1/200 R
68.8	Luton	77.19	pws 15*	77.13	pws 15*	
74.4	Harpenden	84.08		83.26		
79.2	St Albans	89.03	60	88.07	65	
83.8	Radlett	93.07	72	91.57	75	1/176 F
86.5	Elstree	95.56	58	94.30	62	1/200 R
92	Hendon	101.13		99.31		
94	Cricklewood	102.53		101.06	sigs	
97.5	Kentish Town	107.00	sigs	105.16	sigs	
99	St Pancras	110.20	5 ¼ L	108.12	3 ¼ L	
		(106¾ net)		(105 net)		

Mr Charlewood commented that the unrebuilt small-boilered 4-4-0s had no margin on these fast trains for which their permitted load was just 160 tons. By this time, however, the rebuilding of many 4-4-0s with the 'H' boiler, with the increase to 175lbs psi pressure, appeared to give them the edge. By 1908, the thirty 1562 class had all been rebuilt with 'H' boilers and had been renumbered and were allocated as follows:

Derby: 328-336 (1562-1570), 342 (1576)
Manchester: 348-357 (1657-1666)
Skipton: 337-341 (1571-1575), 343-347 (1577-1581)

The ten 1667 class had all been withdrawn by then and replaced by new engines originally allocated the same numbers but renumbered 483-492 in 1907.

Nottingham: 485-489
Leicester: 490-492
Manchester: 483, 484

By the beginning of the First World War, sixteen of the 1562 class had received Deeley G7 boilers. No difference for these sixteen appears to have been made in their shed allocation for the thirty engines were stationed in just two groups, 328-339 at Liverpool Brunswick and 340-357 at Manchester. After the war, the allocation changed little, though 328 returned to Derby, 329 and 340 went to Buxton and 333 to Leicester. No 339 joined the majority of the class that had remained at Manchester.

All were extant at the formation of the LMS in 1923 and retained their Midland 1907 numbers. By this time all Midland express work was in the hands of the Compounds and class 3 and 4 simple 4-4-0s and the class 2s were restricted to secondary stopping passenger

No 1562 class with 'H' boiler, Manchester's 354 (ex 1663) at Buxworth Junction, c1923. (MLS Collection)

One of the class still with the 'H' boiler, 343, topping the Lickey incline at Blackwell with a stopping train from Gloucester, c1925, a year before the engine's withdrawal. Previously a Manchester engine, it was probably based at Saltley or Bristol in its final years. Despite the light load, it seems to have a 'Jinty' 0-6-0T as banker. (Locomotive & General/MLS Collection)

The last survivor of the 1562 class with Deeley G7 boiler and extended smokebox, 338, at Long Eaton with a lightweight express composed of a very motley collection of rolling stock, c1933. (MLS Collection)

No 338 again, on a stopping passenger train at New Mills, 4 August 1935. No 338 was withdrawn in February 1937. (MLS Collection)

No 332 with superheated boiler runs into Manchester London Road with a stopping train from Liverpool Lime Street, summer 1938.
(N. Fields/MLS Collection)

No 40337 at Chinley with a Manchester-Sheffield train, with ex LNWR G2 0-8-0 49416 on a freight, 16 July 1955.
(MLS Collection)

No 1738 class No 1743 built in January 1886, in Works grey after completion. (F. Moore/MLS Collection)

work and the piloting of expresses which remained a feature on the Midland main line for many years after the Grouping until Stanier's Black 5s and Jubilees appeared in any numbers. However, those engines still with the 1907-9 'H' boilers were withdrawn between 1925 and 1928, and those that had the Deeley G7 boilers between 1928 and 1932, with just 342 remaining at Leeds until 1933, and the last one, 338, at Manchester until 1937. The five engines that had been rebuilt as class 483 with superheated boilers in 1923 lasted until the 1950s, 40353 being withdrawn from Wellingborough in July 1953, 40351 from Leeds Holbeck in December, 40356 from Carlisle Upperby in April 1957, 40337 from Hasland in April 1958 and finally 40332 from Bristol (Lawrence Hill) in December 1959 after spending the previous decade at Crewe North until March 1957.

The 1738 class (358-377)
Design & construction
Approval was given for another twenty 4-4-0s in July 1884 and two batches of ten numbered 1738-1757 were delivered from Derby Works between November 1885 and December 1886 authorised at a cost of £2,250 each (actual £2,473). These engines, like the 1667 class, had large 7ft 0½in coupled wheels, reverted to 18in x 26in cylinders and had steel boilers pressed at 160lbs psi, but were otherwise identical to the 1562 class. No 1757, the last of the class, was exhibited at Saltaire (Shipley) in May 1887 at one of Queen Victoria's Jubilee celebrations which was opened by her daughter Princess Beatrice and her husband Prince Henry and it was named *Beatrice* in honour of the occasion. The engine carried the name until it was rebuilt in 1907.

No 1752, built in August 1886, which was reboilered in July 1907 with an H boiler and a G7 in June 1911, surviving in that form until 1940. This photograph was taken c1895. (Real Photographs/ MLS Collection)

The 1887 Jubilee engine 1757 *Beatrice* seen c1900. It was rebuilt with an 'H' boiler in January 1907, and a superheated boiler in 1923, being finally withdrawn, unnamed as 40377, in 1955. (MLS Collection)

No 373, the former 1753, rebuilt with an H boiler in November 1906, and withdrawn still in that form in November 1925. It is seen here shortly after reboilering, c1907. (E.M. Johnson Collection)

Their 'B' boilers were replaced by the Johnson 'H' boilers in 1906 and 1907 and they were renumbered 358-377 in 1907. Ten – nos 358, 361, 363, 365, 366, 368, 369, 372, 374 and 376 – received G7 boilers between 1909 and 1911 and five got the class 483 superheater boiler treatment between 1922 and 1924 (359, 362, 364, 370 and 377 – all engines that did not get the G7 boiler).

Withdrawals started in 1925 and the five still with 'H' boilers had gone by 1927. The ten with G7 boilers lasted longer. Three, 358, 361 and 363, went in 1928 or 1929 but the rest lived on well into the 1930s with two, 369 and 372, not being withdrawn until 1940. The five superheated engines went into BR stock and were withdrawn in the 1950s – 40370 in 1951, 40359 in 1954, 40377 in 1955 and the last two, 40362 and 40364, in 1956.

No 361, built as 1741 in December 1885, rebuilt with an H boiler in May 1907, seen here with the Deeley G7 Belpaire boiler of October 1910, at Derby, c1923. It was withdrawn in December 1929. (F. Moore/ MLS Collection)

No 368, the former 1748 of June 1886, rebuilt with the Deeley G7 boiler, at Kentish Town, June 1935. It was withdrawn in December 1939. (W. Potter/ MLS Collection)

Operations

The initial allocation of the twenty locomotives in 1885/6:

Kentish Town:	1738-1744, 1746, 1747, 1757
Leeds:	1750-1756
Carlisle:	1748, 1749
Bedford:	1745

By 1892 this had changed to:

Kentish Town:	1746, 1757
Leicester:	1738-1741
Bedford:	1742-1745
Nottingham:	1747, 1753-1756
Leeds:	1750-1752
Carlisle:	1748, 1749

Between 1892 and 1902 ten engines were located at Leeds – 1747 to 1756 – and the majority of the rest, 1740, 1742-1746 and 1757, were at Kentish Town. Just three had been based for the first time at Gloucester – 1738, 1739 and 1741. After rebuilding with the larger Johnson 'H' boilers, most of the Leeds engines, 370-376 (1750-1756), moved to Sheffield, 361 and 362 (1741 and 1742) went to Leeds, 363-366 (1743-1746) and the former named 377 (1757) were at Kentish Town, 367-369 (1747-1749) were at Lancaster, 358 (1738) and 360 (1740) at Saltley and 359 (1739) at Bristol.

Cecil J. Allen described a number of runs about the 106 minute schedule from St Pancras to Leicester in the April edition of the 1911 *Railway Magazine*, most with Compounds but one with a 7ft coupled wheel G7 boiler fitted 4-4-0. The smaller engine could not quite keep the booking with the load geared for a class 4. However, in the up direction a class 2 with G7 boiler beat the 105 minute schedule by a minute and a half with the incentive of a late start from Leicester.

St Pancras-Leicester, c1910

7ft 4-4-0 with G7 boiler (no. unknown)

210 tons

Miles	Location	Times	Speeds		Gradients
0	St Pancras	00.00			
1.5	Kentish Town	03.35			1/178 R
6.9	Hendon	11.15			1/200 F
12.4	Elstree	17.55	49		1/176 R
19.9	St Albans	25.20	72		1/176 R
30.2	Luton	36.30		2½ L	
37.3	Harlington	43.10	64		1/200 F
49.8	Bedford	54.00	75	1 L	
59.7	MP 59 ¾ (Sharnbrook)	65.10	35		1/119 R
65	Wellingborough	70.15	70		1/120 F
72	Kettering	77.10			
78.5	Desborough North	85.05	60/36		1/136 R
82.9	Mkt Harborough	90.10	60*		1/132 F
89.7	Kibworth North	96.10	48		1/130 R
95.4	Wigston	102.50			1/199 F
<u>99.1</u>	<u>Leicester</u>	<u>107.00</u>		<u>1 L</u>	

Miles	Location	Leicester-St Pancras, 1910 Johnson 4-4-0 (G7 boiler) 215 tons		
		Times	Speeds	Gradients
0	Leicester	00.00	4½ L	
3.7	Wigston	06.30		
10.2	Kibworth	13.50	40	1/199 R
16	Market Harborough	19.20	66	1/238 F
20.8	Desborough	25.20	40	1/132 R
27	Kettering	30.40	80 5¼ L	1/118 F
34	Wellingborough	36.15	75	
39.3	Sharnbrook summit	42.15	32	1/119 F
49.2	Bedford	50.40	85 3¼ L	
57.3	Ampthill	59.15	49	1/200 R
66.3	Leagrave	70.05	45	1/200 R
68.8	Luton	72.45	4¼ L	
79.2	St Albans	83.20	60	
83.8	Radlett	87.20	78	1/176 F
86.5	Elstree	90.05	42	1/200 R
92	Hendon	95.00	76	
97.5	Kentish Town	100.30		
99.1	St Pancras	103.35	3 L	

As indicated above, the class 2 made a slight recovery of time by very fast running down the long gradients. The Compounds booked for this service did so easily by better hill-climbing, being eased downhill.

During the First World War the entire class was split between Leeds (359-367) and Sheffield (368-377) with just the first of the class, 358, at Manchester Belle Vue. Afterwards

No 1752 moved from Leeds to Sheffield shortly after rebuilding with an 'H' boiler in September 1906 and is seen here shortly before renumbering at Mill Hill on a down express, March 1907. It would be the last of the saturated boiler '1738' class engines withdrawn as 372 from Bourneville in 1940. (E. Pouteau/MLS Collection)

No 360 of Leeds works a Carlisle-Leeds special train over the water troughs at Garsdale, February 1922.
(F.E. Mackay/MLS Collection)

No 40377, the former 1757 *Beatrice,* now superheated as a class '483' 2P, runs into Timperley station with the 3.45pm Manchester Central-Chester on the Cheshire Lines Committee (CLC) route, 8 October 1952.
(R.E. Gee/MLS Collection)

the Leeds engines remained there, but the Sheffield engines were dispersed with, just 369, 373, 374, 376 and 377 staying. No 368 was allocated to Gloucester, 371 to Worcester, 375 was allocated to Saltley, and 372 joined its sisters at Leeds. No 358 remained at Belle Vue. A couple of engines – 365 and 372 – had short spells at Bedford in the late 1920s.

Apart from the five superheated engines rebuilt to the '483' class, only engines with G7 Belpaire boilers were left in the 1930s with 365 at Leeds, 368 and 369 at Saltley, 372 at Hasland and 374 and 376 at Burton. By this time, their work was mainly stopping passenger trains and the piloting of some expresses when loads exceeded those stipulated for the Compounds. The last G7 boiler fitted 1738 class, 369, was shedded at Birmingham Bourneville when it was withdrawn in 1940.

The superheated engines joined the other 4-4-0s similarly rebuilt or built new as '483' class and passed to British Railways in 1948. No 40370 expired quickly from Toton in 1950, but 40359 remained active at Hasland until 1954, 40377 (the former *Beatrice*) on the North Wales lines at Rhyl, Chester and Llandudno Junction until September 1955, 40364 at Burton-on-Trent to June 1956 and the final survivor, 40362, at Lancaster until December 1956.

The 1808 class (378-402)
Design & construction
The continued expansion of the Midland Railway services put pressure on the company's motive power fleet and ten additional 4-4-0s were authorised in February 1887. Nos 1808-1817 were constructed at Derby and were delivered between April and June 1888. A further order for five was made in January 1888 and 1818-1822 were delivered from Derby Works by September of that year. Then, to replace older engines withdrawn from stock, ten more were authorised in May 1890 from the Revenue instead of the Capital account and although to the same technical specification and dimensions, were numbered 80-87, 11 and 14 filling gaps in the company's number system. The main dimension change for this series was yet another variation of coupled wheel diameter, reduced to 6ft 6in. Cylinders remained at 18in x 26in and boiler pressure of 160lbs psi had now become standard.

No 1814, as built in 1888, at Newton Heath depot, c1895. (E.M. Johnson Collection)

No 1819 (later 389) as constructed in 1888 and seen at Newton Heath, Manchester, c1900. It was rebuilt with a larger 'H' boiler in 1905.
(F. Moore/MLS Collection)

No 14, the last of the 1808 class built in August 1891, rebuilt with 'H' boiler in October 1904, at Kentish Town, c1905. (Locomotive Publishing Co./MLS Collection)

No 392, formerly 1822, rebuilt with 'H' boiler in June 1904, at Manchester Belle Vue, c1910. It was rebuilt with a G7 boiler in January 1912. (W.L. Good/MLS Collection)

No 400(ex 87) built in June 1891, rebuilt with an H boiler in June 1904 and seen here at Manchester Belle Vue depot, c1920. It would be replaced by a superheated '483' class version in October 1922 and be withdrawn in January 1949.
(MLS Collection)

No 378, formerly 1808, the first of the series, rebuilt with a G7 boiler in November 1911, seen here at Derby, 22 August 1937. (N. Fields/ MLS Collection)

The initial order for the Johnson larger boilers, type 'H', was made in November 1903 and intended for the 6ft 6in 4-4-0s, so the 1808 class was the first to be rebuilt and all twenty-five received them in 1904 and 1905. The initial order of H boilers were pressed to 175lbs psi. The rebuilt engines were renumbered 378-402 in the Midland 1907 scheme.

In 1908, 386 (the former 1816) was rebuilt with new frames and a Belpaire boiler and as a result an order was given later in the year to replace the H boilers by the G7 Belpaire boiler on the 1562, 1738 and 1808 classes. The first sixteen '1808s' (378-392) were rebuilt therefore between 1909 and 1912 when the new superheated boiler was developed for the '483' class. As a result, the rebuilding of the 1808s and earlier classes ceased as new engines with the superheated boilers were constructed and it was not until the 1920s that those still with H boilers were rebuilt with superheated rather than G7 boilers. Nos 394-397 and 400-402 were rebuilt in 1922 and 1923. Nos 398 and 399 (the former 85 and 86) retained their H boilers until their early withdrawal in 1925 and 1926.

The first G7 boilered member of the class to be withdrawn was 393 (ex 80) in December 1928, but most lasted well into the 1930s, with 378 (the prototype 1808) lasting until October 1947, and three surviving nationalisation in 1948. The renumbered 40385 and 40391 were both withdrawn in September 1949 and 40383 (ex 1813) was not withdrawn until July 1952. Five of the seven of the class with superheated boilers, identified in the previous paragraph, were retained into the BR era. Nos 394 and 400 were the only ones that were withdrawn earlier, 40397 being withdrawn in 1951, 40401 in 1952, 40395 in 1954, 40402 in 1960 and 40396 the last, surviving well into the 1960s.

The Midland Class 2 • 49

No 40397, one of the seven 1808 class rebuilt with superheated boilers in 1922/3, at Springs Branch shed, Wigan, 24 April 1949. (B.K.B. Green/MLS Collection)

No 40383, the last of the 1808 class retaining a G7 boiler, seen here at Derby in lined mixed traffic BR livery, c1951. It was withdrawn in July 1952. Note the tall chimney retained on this engine. (MLS Collection)

Operations

The justification for the latest batch of class 2s was the expansion of Midland Railway services via the Lancashire and Yorkshire Railway from Hellifield to Manchester, Liverpool, and Marple. Then new services on the Skipton-Ilkley line proposed in 1888 set a requirement for additional engines justifying the building of five more of the class. Therefore, the initial allocation made to meet these needs was:

Manchester:	1808-1812
Liverpool:	1813-1818
Hellifield:	1819-1822, 80-83
Carlisle:	84-87
Nottingham:	11, 14

The Manchester engines, although allocated to Belle Vue shed, were often stabled at the L&Y shed at Newton Heath when working from Manchester Victoria to Blackburn and Hellifield and the Liverpool engines often used the L&Y Sandhills depot for a similar reason. Reallocations soon took place and the allocation in 1892 was:

Hellifield:	1808-1813, 80-83
Newton Heath:	1818-1820
Lower Darwen:	1821, 1822
Belle Vue:	1814, 1815
Heaton Mersey:	1816, 1817
Leeds:	11, 14
Carlisle:	84-87

After delivery of some of the larger boilered 4-4-0s in the 1890s, the Carlisle four were dispersed to Skipton and Hellifield. Then again, after the rebuilding with H boilers in 1904/5, and the 1907 renumbering, the allocation was:

Manchester:	380-392 (ex 1810-1822)
Liverpool:	378, 379 (ex 1808, 1809)
Skipton:	393-400 (ex 80-87)
Nottingham:	401, 402 (ex 11, 14)

The First World War brought about some changes with two Skipton and the two Nottingham engines still with 'H' boilers, 399-402, moving to York to replace 2-4-0s on cross country workings from the North East to the West Midlands. The Midland route through Birmingham was getting the class 2 'H' boilered 4-4-0s as the superheated and G7 boilered 4-4-0s covered the main

A pair of Manchester based 'H' boiler fitted 1808s, 381 and 383, on the 4.55pm Hellifield to Manchester Victoria, c1908. (MLS Collection)

G7 boilered 381 of Manchester Belle Vue with a Manchester-Liverpool train near Hunts Cross, c1912.
(F. Moore/MLS Collection)

Belle Vue's 386 departs from Manchester Victoria for Hellifield, an LNWR 'coal tank' in the background, c1910. No 386 was rebuilt with a G7 boiler in May 1909.
(G.W. Smith/MLS Collection)

line Midland expresses, and before the Grouping 395 was at Saltley and 397 at Worcester.

After seven 'H' boilered 1808s were rebuilt with superheated boilers at the time of the Grouping, 378-393 remained around Sheffield in the 1920s, although four of them – 381 and 391-393 – were transferred to Carlisle Upperby for working the Cumberland coast route. No 381 returned to Sheffield in 1933, but two of the others had been withdrawn.

The G7 boilered survivors were allocated as follows in September 1933:

Burton:	378
Sheffield:	379, 381-383, 385, 387, 389
Carlisle (Upperby):	391

Belle Vue's 384 at the head of an express for Hellifield at Manchester Victoria, c1911. (G.W. Smith/MLS Collection)

No 383 leaving York with a stopping train for Leeds and Sheffield, c1922. No 383 received the G7 boiler in December 1909 and was the last to remain in traffic until 1952 in this condition. (Real Photographs/MLS Collection)

Sheffield's 391 pilots a Midland 3P No 725 out of Derby with a train for Nottingham, c1922. (MLS Collection)

Sheffield's 382 entering Totley Tunnel with a Sheffield-Manchester train, c1932. (J. Maynard Tomlinson/MLS Collection)

Four 1808s with G7 boilers were retained at the onset of the Second World War – 378, 383, 385 and 391, with 378 being withdrawn in 1947. No 40383 was given a general repair at Derby in June 1948 and it was based at Derby, often used on inspection saloon specials. It was the last of the saturated Midland 4-4-0s and it was withdrawn in July 1952 with a recorded mileage of 1,604,149.

Five of the seven 'H' boilered 1808s that were given type '483' superheated boilers in 1922/3 were taken into BR ownership in 1948 and were withdrawn as follows:

40395 from Nottingham in 9/54 (having previously been several years at Burton)

40396 from Burton in 2/61

40397 from Liverpool Brunswick in 3/51

40401 from Spital Bridge in 10/52

40402 from Leicester in 9/60.

No 397, with superheater boiler from October 1922, leaving Burton with a down express, 9 May 1925. (W.L. Good/MLS Collection)

Superheated 395 leaving Wickwar Tunnel with a Birmingham-Bournemouth train routed via the Somerset & Dorset line via Mangotsfield and Bath, 1935. (Locomotive & General/MLS Collection)

No 397 piloting 523 (Midland '60' class also rebuilt with superheated boiler) on a down express between King's Norton and Northfield, Birmingham, 1936. (W.L. Good/MLS Collection)

No 385, still with a G7 boiler, on a Manchester-Sheffield stopping train at Chinley, 17 May 1947. No 385 was one of the three saturated class 1808s that became BR property. It was withdrawn in September 1949. (MLS Collection)

No 40396 leaving Farnworth with the 12.30pm Liverpool-Manchester, 6 June 1949. This engine, rebuilt and superheated in May 1923, was the last survivor of the 1808 class, and was still operating out of Burton-on-Trent until February 1961. (R.E. Gee/MLS Collection)

The 2183 class (403-427)
Design & construction

Johnson once more reported to the Midland Railway Board in October 1890 that the company had insufficient large passenger engines and persuaded them to authorise a further twenty 4-4-0s with 7ft coupled wheels. Sharp, Stewart's tender of £2,520 was accepted and 2183-2202 were delivered between April and July 1892. These were equipped with Johnson's 'D' rather than the smaller 'B' boiler – longer and pressured at 160lbs psi. The cylinders were also enlarged slightly to 18½in diameter and they were fitted with slide valves. Five more with 7ft wheels were ordered in July 1895 and delivered in August and September 1896, numbered 156-160. The main dimension changes to earlier 4-4-0s were:

Cylinders (2 inside):	18 ½in x 26in
Coupled wheel diameter:	7ft 0in
Boiler pressure:	160lbs psi
Heating surface:	1,205sq ft
Grate area	19.6sq ft
Water capacity:	3,250 gallons
Coal capacity:	4 tons

No 2183 class No 2202 constructed by the Sharp, Stewart Company in July 1892, as built in Works grey livery. It was rebuilt with an H boiler in January 1906 and a superheated '483' boiler in April 1922. Renumbered 422 in 1907, it was eventually withdrawn from Skipton as 40422 in October 1953. (F. Moore/MLS Collection)

No 2193 in Midland red livery, at Bedford, c1900. It was built in May 1892, renumbered 413 and fitted with an H boiler in 1907 and rebuilt with a superheated boiler in September 1918, being finally withdrawn as 40413 from Kentish Town in August 1958. (Loco Publishing Co./MLS Collection)

No 2201 built in July 1892, seen here at Derby, c1895. (E.M. Johnson Collection)

No 2202 built in July 1892, seen here at Derby from the other side, c1895. (E.M. Johnson Collection)

No 406, built in 1892 as 2186, and rebuilt with an H boiler in May 1906, seen at Manchester Central, c1912. It was replaced with a superheated 483 class version in 1914.
(E.M. Johnson Collection)

Nottingham's 403, the former 2183, rebuilt as '483' superheated class in December 1920 and oil-fired during the coal strike of 1921, at Attenborough.
(MLS Collection)

No 2183 class 40418, formerly 2198, built by Sharp, Stewart in July 1892, fitted with H boiler in July 1906 and '483' superheated boiler in June 1914. It was photographed at Derby on 28 August 1954 and was withdrawn from Derby shed in August 1956. (MLS Collection)

No 2183 class 40426 (formerly 159) in more typical 1950s condition at Gloucester, 19 April 1954. It was built as 159 in September 1896, equipped with an H boiler in 1906 and a superheated '483' boiler in June 1916. It was withdrawn from Bristol Barrow Road depot in November 1957. Note the external exhaust steam injector pipe.
(A.C. Gilbert/MLS Collection)

All received the lined red Midland Railway livery with crest on the leading splashers and the initials MR on the tender. The class was renumbered 403-427 in 1907. The twenty-five engines were rebuilt with 'H' boilers between January 1906 and April 1908. None received the G7 Belpaire boiler, but all were rebuilt to the 483 superheated boiler version between 1914 and 1922. All except 403, 408 and 427 lasted in superheated form until the 1950s, the BR numbered 40411 and 40421 not being withdrawn until the 1960s.

Operations

The initial allocation of the twenty locomotives of this series built in 1892 was:

Nottingham:	2183-2192
Kentish Town:	2193-2197
Bedford:	2198-2202

The five additional engines of the class built in 1896 were all allocated to Nottingham which was the main base of the 2183 series. By the turn of the century, there had been very little change, apart from the Kentish Town five engines, 2193-2196, being transferred to Saltley and 2197 to Gloucester.

After rebuilding with the 'H' boiler, 2197 joined her sisters at Saltley, two of the Nottingham engines, 2191 and 2192, went to Sheffield and two of the Bedford engines, 2198 and 2199, were transferred to Kettering, the first

No 2195, built in May 1892, pilots a 2-4-0 on a down express at Hendon, c1895. (Loco Publishing Co./ MLS Collection)

No 2184 heads a St Pancras-Nottingham express near Mill Hill, c1903. (Locomotive & General/MLS Collection)

No 414, rebuilt with the H boiler in April 1904, at Manchester Central with a southbound express, c1910. It would be 'replaced' as a superheated '483' class in July 1914. (MLS Collection)

allocation of any of the 4-4-0s to this shed.

160, allocated new to Nottingham when built in September 1896 and still there in 1906 after rebuilding with an 'H' boiler, headed the 12.45pm Leeds from Nottingham shortly afterwards and before renumbering in 1907. It was booked to run the 72 miles from Kettering to London in 77 minutes and completed the run in 73 minutes 50 seconds. It hauled the normal load of 195 tons and did not fall below 44mph on the 1 in 120 climb to Sharnbrook summit, with high speeds of 78½ on the descent at Bedford and 76½ near Radlett. Then 2188, rebuilt with an H boiler in 1906 and renumbered 408 in 1907, was recorded between Derby and Manchester around 1908, details in the table below.

No 426 of Nottingham, renewed in June 1916 with superheated boiler, at Northfield with a substantial load with stopping train headcode, 1921. (W.L. Good/MLS Collection)

Derby-Manchester, c1908
10.25am St Pancras-Manchester
408
187/200 tons

Miles	Location	Times	Speeds		Gradients
0	Derby	00.00			
5.3	Duffield	08.35			
10.4	Ambergate	14.45	sigs	¼ E	
17.2	Matlock	23.25			1/177 R
21.6	Rowsley	28.10		¾ E	
25	Bakewell	32.45			1/102 R
27.3	Longstone	36.50			1/100 R
31.4	Millers Dale	43.20		¾ E	1/100 R
36	Peak Forest	51.40	35	1¼ E	1/90 R
41.6	Chinley	58.10		¾ E	1/90 F
44.3	New Mills S. Jcn	60.55		1 E	1/89 F
53.4	Cheadle Heath	69.55	66	1 E	1/100 F
59.2	Chorlton	74.45	sigs	3¼ E	
61.4	Manchester	82.00	(80¾ net)	3 E	

Around 1912, the Midland Railway authorities decided to rationalise the allocation of the class 2 4-4-0s as there was now a substantial number, and the entire class of twenty-five locomotives was concentrated at Nottingham. By 1914, a start had been made on total rebuilding of the engines with the class '483' superheated boilers and in many cases frame renewals and other substantial repairs, so that they were in effect replacements and virtually new engines. Whereas for some of the earlier series only a small number of locomotives were selected for this substantial rebuilding, the 2183 class went straight from H boilered engines to the superheater version without the intermediary rebuilding with the Belpaire G7 boiler.

The 'replacements' went back to the same depots, so Nottingham virtually retained its complete allocation, with twenty-one of the twenty-five still there at the Grouping in 1923. Sheffield had regained 412 and 413 (the former 2191 and 2192 that had been at Sheffield in 1908) and 414 (ex 2194) and 420 (ex 2200) were based at Gloucester.

The first major changes took place in the 1930s when the influx of Stanier 4-6-0s led to the withdrawal of many of the former pre-grouping company express engines, particularly those of the LNWR and L&Y. This led in turn to the need for lower powered engines on secondary routes previously covered by these, and many ex-Midland 4-4-0s, replaced on the main St Pancras expresses by the Stanier 'Black 5s' and 'Jubilees', were reallocated over the system. Sheds which received the former Midland 2Ps as they were now classified included Camden, Northampton, Walsall, Bushbury, Chester and Llandudno Junction. Buxton had acquired 403, 412 and 413 by this time and 405 was at Birkenhead in the late 1930s, being transferred to Abergavenny in October 1939. At the onset of the Second World War further reallocations took place.

The whole class was taken into BR ownership in January 1948 with the first withdrawal, 408, in December of that year. No 427 was the next to go in May 1950 and then withdrawal took place gradually throughout the 1950s with two lasting into the 1960s – 40421 being withdrawn in January 1961 and the last survivor, 40411, in February, both from Nottingham where they had started their lives in a very different form nearly seventy years previously. During the 1950s, the 2Ps were allocated to a variety of sheds, the larger allocations being at Nottingham (6), Derby (6) Northampton (5) and Crewe North (4) at some period. Other sheds that received one or two at some time in the '50s included Heaton Mersey, Stafford, Nuneaton, Rhyl, Hasland, Sheffield, Normanton, Mansfield, Leicester, Kentish Town, Spital Bridge, Carlisle Upperby, Preston, Skipton, Bristol Barrow Road and

No 403 (the former 2183) pilots rebuilt 'Royal Scot' 46157 *The Royal Artilleryman* on a 15-coach up express approaching Shap summit, 27 August 1949. (J.D. Darby/MLS Collection)

Hasland's 409 (the former 2189) working home by piloting Midland Compound 1021 on a Manchester-Sheffield train at Chinley North, 14 May 1949. (J.D. Darby/MLS Collection)

No 40411, the last survivor of the series, at Chinley on a stopping train to Sheffield, 20 June 1953. No 40411 still has eight years to live before withdrawal from Nottingham shed in 1961. (J.D. Darby/MLS Collection)

No 40411 of Nottingham piloting Jubilee 45597 *Barbados* on the up *Thames-Clyde Express* (9.20am Glasgow St Enoch) at Clay Cross, 11 July 1959. No 45597 is showing signs of being short of steam and needing the 2P's assistance on this 11-coach train. (A.C. Gilbert/MLS Collection)

Gloucester. In 1957, when some of the main Midland line expresses were accelerated, a number of 2Ps were allocated to Kentish Town and Derby for piloting the Black 5s and Jubilees on these services when train strengthening took place. This included 40412 which was based at Derby and 40413 at Kentish Town.

The 2203 class (428-472)
Design & construction

The Midland Railway system continued to grow and in December 1891 Johnson sought an additional fifteen 4-4-0s at an estimated cost of £2,500 each for the new line being constructed from Manchester to Sheffield via the Dore and Chinley route. The lowest tender was again from the Sharp, Stewart Co., but at £2,800, £300 in excess of the estimate. Nos 2203-2217 were delivered between January and April 1893. The only significant difference from the 2183 series also built by Sharp, Stewart, was reversion to 6ft 6in coupled wheel diameter instead of 7ft. They had the same 18 ½in x 26in cylinders, slide valves and the enlarged 'D' boiler at 160lbs psi. The total heating surface remained at 1,205sq ft. They were provided with the larger 3,250 gallon tenders.

The replacement of earlier Kirtley engines was becoming necessary and between February 1894 and December 1895 a further thirty were constructed to the same design, but constructed at Derby Works and numbered 184-199, 161-164 and 230-239 in that order, all filling numerical gaps as these engines were financed from the revenue rather than capital account.

The Midland Class 2 • 67

No 233 of the revenue account batch of the 2203 class, built in October 1895, at Leeds, c1900. (Locomotive Publishing Co./MLS Collection)

No 235 built in November 1895 and allocated to Leeds for operation over the Settle & Carlisle line, rebuilt in October 1904 with an 'H' boiler, here c1905. (Locomotive Publishing Co./MLS Collection)

No 2209 shortly after rebuilding with the 'H' boiler in April 1904, at York, c1905. Note that this was one of the first locomotives to gain the numberplate fixed to the smokebox door. (Locomotive Publishing Co./ MLS Collection)

No 428, the prototype 2203, rebuilt with an 'H' boiler in May 1905, renumbered in 1907 and withdrawn, still with this type of boiler in April 1927, at Bedford, 6 August 1923. (A.G. Ellis/ MLS Collection)

Three 'H' boilered 2203 class 4-4-0s were 'replaced' in December 1914, 464, 466 and, illustrated here, 437. The new superheated boilers and frames are complete and awaiting other parts from the withdrawn engines for assembly at Derby Works. This is an important photo as it shows the extent of the rebuilding of the so-called 'replacements' of the Johnson 4-4-0s. (MLS Collection)

No 447, rebuilt to '483' class in March 1920, at Buxton shed, March 1936. No 40447 was withdrawn from Nottingham shed in May 1958. (W. Potter/MLS Collection)

M436, rebuilt to '483' class in September 1914, at Derby, and renumbered with the prefix 'M' immediately after nationalisation, at Derby station, c1948. No 40436 was withdrawn from Burton shed in September 1954. (MLS Collection)

No 40443, rebuilt with superheated boiler and new frames in January 1915, newly renumbered and repainted in BR mixed traffic lined livery at Crewe with 46206 *Princess Marie Louise*, 1949. No 40443 survived to September 1960 after years at Stafford, then in the late 1950s at Saltley, spending its last year at Toton. (F.F. Moss/ MLS Collection)

No 40454, rebuilt as class '483' in May 1922, at Derby, 21 April 1954. It was withdrawn from Nottingham shed in September 1960. (MLS Collection)

The 2203 series was the first group of locomotives authorised for rebuilding with the 'H' boiler. No 2205 was the first to be rebuilt in June 1904, four more were converted later that year and the complete class was equipped by May 1906. In 1907, the whole class was renumbered 428-472. There were visual differences – especially noticeable was the smokebox door – for example 184, 199, 235 and 2217 had a central locking wheel, while others including 163, 2209 and 2216 had flat smokebox doors with front numberplates that later became standard for Midland and subsequently LMS locomotives. Deeley also redesigned the cab to give a little more protection to the crews, especially valuable on the rainy and windswept Settle and Carlisle route.

Just two of the series were subsequently rebuilt with the Belpaire G7 boiler in 1910 -460 and 465. Thirty of the series were 'rebuilt' or 'replaced' by the '483' superheated boiler engines between September 1914 (436) and November 1923 (450 and 472). The unrebuilt engines – 428, 429, 431, 435, 440-442, 445, 449, 451, 457, 459, 465, 467 and 469 (15 engines) were mostly withdrawn between March 1925 (449) and September 1927 (435), with just 460 and 465 (the two G7 boilered engines) lasting until February 1930 and November 1931 respectively.

The superheated 'replacement' engines were all taken into BR ownership in January 1948, the first withdrawals being 437, 456 and 466 in 1949 and 446 and 468 in 1950 before renumbering with the prefix '40'. The final Works repair was 40452 in February 1958. The last survivors were 40454 withdrawn in September 1960 and 40439, 40443 and 40452 in January 1961.

Operations

The 1893 built locomotives were divided between Leeds and Manchester Belle Vue depots with the first four, 2203-6, at Leeds, the rest at Belle Vue. The first four were delivered before the Dore and Chinley route was opened, so initially were used on the Settle & Carlisle line and they were joined at the northern end at Carlisle in 1895 by 184-199, the 6ft 6in engines replacing the earlier 7ft examples on this heavily graded route. Nos 161-164 were allocated to Leicester and 230-235 joined the earlier locomotives of the class at Leeds. The final four of the batch broke new ground, being allocated to Carnforth.

A major shift took place at the beginning of the twentieth century, with the first four Leeds engines and 2207-2213 of Manchester going to Sheffield. Four other Belle Vue engines, 2214-2217, were dispersed to sub-sheds in the Manchester area and two of the Carnforth engines – 236 and 237 – moved to Skipton. After rebuilding with 'H' boilers and renumbering, the allocation in 1908 was as follows:

Sheffield:	432-442 (2207-2217), 451 (192), 457 (198), 459-462 (161-164)
Carlisle:	443-450 (184-191), 452-456 (193-197), 458 (199), 467 & 468 (234 & 235)
Liverpool Brunswick:	428-431 (2203-2206)
Leeds:	463-465 (230-232)
Skipton:	469-472 (236-239)
Derby:	466 (233)

By the beginning of the First World War, the entire class was based in the north, 428-447 at Carlisle and 448-472 at Skipton. Most were rebuilt/replaced with superheated boiler versions between 1914 and 1923. The unrebuilt engines were allocated between 1921 and their withdrawal in the late 1920s as follows:

Carlisle:	438, 442, 445
Skipton:	460 (G7 boiler), 465 (G7 boiler), 467, 469, 471, 472
Leeds:	440, 441
Lancaster:	449, 451, 454
Nottingham:	428, 450, 457
Saltley:	429, 431, 435, 439
Gloucester:	459

One of the unrebuilt Saltley allocation around the Grouping period in the early 1920s was recorded in a 1924 *Railway Magazine* still on express work, though class 2s, especially those still with the 'H' boilers, would soon be replaced by LMS built Compounds and then Stanier's 4-6-0s on the Leeds-Birmingham-Bristol main line.

No 2209 at Sheffield with the royal train, 21 May 1897. (E.M. Johnson Collection)

No 2206 on the route for which it was built, at Dore & Totley station with a Sheffield-Manchester train, c1900. (MLS Collection)

No 238, rebuilt in December 1904 with an 'H' boiler, and allocated to Skipton, at Hellifield station, 13 June 1905. (G. Waite/MLS Collection)

No 430, formerly 2205, rebuilt in June 1904 with an 'H' boiler, and newly allocated to Liverpool Brunswick, at Halewood on the Cheshire Lines Committee system, c1908. (F. Moore/MLS Collection)

Carlisle's pair, 432 piloting 433, leaving Carlisle with an express for Leeds via the Settle & Carlisle line, c1912. (MLS Collection)

Miles	Location	Cheltenham-Birmingham New Street 431 (non-superheater) 240/260 tons (Special limit 180 tons) 7.20pm Bristol Mail			
		Times	Speeds		Gradients
0	Cheltenham	00.00			
3.8	Cleeve	06.05			
7.2	Ashchurch	09.30	64		1/295 F
9.4	Bredon	11.50			1/301 R
13.2	Defford	15.35	59	L	
18	Abbots Wood Jcn	21.15	50½		1/301 R
20.4	Spetchley	24.05			
24.6	Dunhampstead	28.45	57	L	
31.3	Bromsgrove	36.45			
0		00.00 Attach 'Big Bertha' 0-10-0			
2.2	Blackwell	07.40	17 ave.		1/37¾ R
3.6	Barnt Green	10.20	31		
8.7	King's Norton	16.20	51/60		1/301 F
10.9	Selly Oak	20.00			
14.2	Birmingham	25.15			

No 431 (ex 2206) of Saltley depot climbing the Lickey bank, approaching Blackwell, with a Bristol-Derby express, c1920. No 431 was withdrawn in this form in August 1925. (Locomotive & General/MLS Collection)

No 463 (ex 230), in '483' superheated form, piloting ex LNWR 'Claughton' 4-6-0 5932 out of Carlisle Citadel with an express for Leeds via the Settle & Carlisle line, c1925. (MLS Collection)

No 446 (ex 188) at Chinley East Junction with a Hope Valley train, 1934. (E.R. Morten/MLS Collection)

No 462 (ex 164) at New Mills South Junction with a Manchester-Buxton train, 1 August 1936. (MLS Collection)

No 461 (ex 163) with a Manchester-Sheffield train at Chinley, 29 May 1939. (MLS Collection)

78 • MIDLAND RAILWAY AND L M S 4-4-0 LOCOMOTIVES

No 438, one of the last of the class rebuilt with superheated boiler and new frames in 1923, was based at Carlisle for many years, but is here seen leaving New Mills Tunnel with a Manchester-Sheffield local train, 7 April 1947. It was withdrawn from Patricroft in 1954. (MLS Collection)

The superheated engines, like the rebuilt 2183s, were spread all over the LMS system in the late 1920s and 1930s for secondary passenger and pilot engine work. In BR days, by 1950, the allocation had settled as:

Chester:	40430
Rhyl:	40433
Burton:	40432, 40436, 40453
Patricroft:	40434, 40450
Nuneaton:	40438, 40447
Bourneville:	40439, 40463
Stafford:	40443, 40461, 40471
Royston:	40444
Carlisle Upperby:	40448
Nottingham:	40452, 40458
Mansfield:	40454
Manningham:	40455
Walsall:	40462
Liverpool Brunswick:	40464
Hellifield:	40470
Hasland:	40472

Pilot working of the accelerated Midland main line expresses was

rife between 1957 and 1959, the last of this class involved regularly in such turns being 40452 of Nottingham. For the last depot of each member of the class, see the statistics in the appendix, pages 312-3. The final 1960-1 withdrawals were 40439 and 40443 from Toton, 40452 from Leicester, 40453 from Burton and 40454 from Nottingham.

The 2581 class (473-482)
Design & construction
Twenty 4-4-0s were ordered from Beyer, Peacock in March 1899 at a cost of £3,230 each, half for the Midland & Great Northern Railway (see page 120). The company offered a similar design to the 1891 class 1808 locomotives, albeit with 18½in rather than 18in cylinders. A retrograde step was the equipping of these

No 40464 of Liverpool Brunswick (transferred to Northampton in 1951), departs from Baguley station with the 3.53pm Stockport-Glazebrook local train, 7 October 1950. (MLS Collection)

No 2587, built in April 1900, at St Pancras, c1901. It was rebuilt with an 'H' boiler in 1905, renumbered 479 in 1907 and superheated with new frame in March 1917. (MLS Collection)

No 2587, rebuilt with 'H' boiler and smokebox door with spoked ring and dart in March 1905, at York, c1912. (F. Moore/ MLS Collection)

engines with the smaller 'B' rather than 'D' boiler. They had 6ft 6in coupled wheels and were numbered 2581-2590. However, like the 1808s, they were pressed at 160lbs psi. They were delivered between March and May 1900. The ten locomotives were rebuilt with 'H' boilers between December 1904 and August 1905 and renumbered 473-482 in 1907.

Five were rebuilt as class '483' superheated engines with 7ft coupled wheels, 482 in September 1914, 479 in March 1917 and 477, 478 and 480 in 1922. The other five retained their 'H' boilers until withdrawn between October 1925 and August 1927 (476).

Operations
All ten locomotives were initially allocated to Derby. Their allocation after boiler rebuilding by 1908 was:

Leicester:	473, 474
Leeds:	475, 476
Derby:	477-482

Four of the five 7ft replacements lasted into the 1950s and were withdrawn as follows – 40478 from Nottingham in 1950, 40477 from Kentish Town in 1951, 40480 from Normanton in 1954 and finally 40482 from Sheffield Millhouses in July 1957, after spending the previous four years at Longsight.

No 476, the former 2584, rebuilt with an 'H' boiler in February 1905, at Leeds Holbeck depot, 11 September 1926. It was withdrawn without being rebuilt with a superheated boiler in June 1927. (W. Potter/MLS Collection)

No 2590, built in May 1900 and rebuilt with an 'H' boiler in August 1905, at Mill Hill on a down express, c1906. It was renumbered 482 in 1907 and rebuilt as a superheated '483' class in 1914. (MLS Collection)

Derby's 'H' boilered 473 (ex 2581) after arrival at Bristol Temple Meads from Derby and Birmingham, c1907. It was withdrawn without further rebuilding in June 1927. (Locomotive & General/MLS Collection)

The 1667 replacement class (483-492)

Design & construction

The less-than-successful Joy valve gear 1667 class were withdrawn and replaced with similarly numbered engines with new boilers, frames, piston valves and motion – in effect new engines. The five in poorest condition were renewed first between October 1896 and March 1897 – 1667, 1668, 1672, 1675 and 1676. Two more were 'rebuilt' in 1897 – 1669 and 1671 – and the last three in June 1901. Like the other Johnson 4-4-0s, they were reboilered with 'H' boilers between 1906 and March 1908 (1676 was the last). They were renumbered 483-492 in 1907. They retained the 19in diameter cylinders of the original engines although 1671 and 1674-1676 had their cylinders lined up to 18½in later. They had 'D' boilers pressed at 160lbs psi and 1670 and 1672 had piston valves. Their 2,950 gallon tenders were replaced with larger 3,250 gallon ones.

In 1912, it was decided to fit the class 2 4-4-0s with superheated boilers and despite their relatively recent construction, the replaced 1667s became replaced in turn by what became known as the '483' class, all ten being rebuilt in 1912 and 1913. They were considered rebuilds for accountancy processes, but probably only the wheels and other small parts were incorporated. Royalties for the superheater design were cheaper for rebuilt engines which may signify why they were so classified. Dimensions for the rebuilt '483' class were:

Cylinder diameter	20½in x 26in
Piston valves	8in diameter
Coupled wheel diameter	7ft 0½in
Bogie wheel diameter	3ft 6½in
Boiler pressure	160lbs psi
Heating surface	1,483sq ft (of which superheater was 313sq ft) (Later 1,410sq ft with superheater 253sq ft)
Grate area	21.1sq ft
Axleload	17½ tons
Weight – Engine	53 tons 7 cwt
– Tender	39 tons 16 cwt
– Total	93 tons 3 cwt
Water capacity	3,250 gallons
Coal capacity	4 tons
Tractive effort	16,551lbs

The first replacement of 1667 constructed in March 1897 and rebuilt with the 'H' boiler and renumbered 483 in July 1907. It is seen here shortly afterwards and before being replaced by the '483' class of superheated 4-4-0s that only used a few parts of the original engine.
(E.M. Johnson Collection)

No 483, the prototype of all the rebuilt superheated new framed class 2 4-4-0s, later LMS 2P, most of which lasted well into BR ownership. It was built in November 1912 and is seen here in Works Grey for official photographic purposes. It was not, however, the first rebuild which was 488 in March of that year.
(MLS Collection)

No 488, shortly after its March 1912 rebuilding from the earlier 1672 (itself a replacement for the Joy valve gear engine of the same number) at Manchester Central, c1912. No 488 was the first engine to be rebuilt as a '483' class superheated engine. (E.M. Johnson Collection)

No 40491, rebuilt in May 1913 to replace 491 (ex 1675), which itself had replaced the earlier Joy valve gear 1675, at Nottingham in BR mixed traffic lined black livery, 9 July 1950. (J.D. Darby/ MLS Collection)

Operations

The first set of renewed engines went back to the depots from which their predecessors had been withdrawn, the last known allocation being six at Nottingham, two at Derby and two (1675 & 1676) at Kentish Town. The allocation in 1902 of the replacements was:

Manchester Belle Vue:	1667-1669
Nottingham:	1670, 1671, 1673, 1674
Kentish Town:	1675, 1676

The whereabouts of 1672 was not recorded but is believed to have been Nottingham also. By 1908, 1669 had been transferred to Nottingham and 1674-1676 were at Leicester.

The London-Leicester booked maximum load for a superheated class 2 of the '483' series was 200 tons, only 40 less than the Compounds, though the increase in loading during the First World War and its aftermath meant that double-heading was frequent. Often the motive power for the principal expresses was a 4-4-0 and a Johnson 4-2-2 as pilot, though a pair of 4-4-0s was common. 489 (the former May 1901 replacement for 1673), rebuilt in March 1908 with an 'H' boiler, was piloted in 1911 on a Nottingham-London train by Johnson 4-2-2 'Single' No 677 throughout, with a heavy load for the Midland expresses, 275 tons. The pair left Nottingham 7 ½ minutes late, achieved 54mph on the 1 in 200 past Plumtree and cleared Widmerpool as the gradient eased at 60. The schedule was tight, however, and they exceeded the 22 minutes allowed for the 18¼ miles mainly against the grain by just a quarter of a minute. They then attained 60mph on the level before Oakham and tore down the 1 in 142 to Manton at a full 80mph. They averaged 60mph up the 1 in 200 to Corby and came to a stand in Kettering in 35 minutes 50 seconds for the 33¼ miles, having regained nearly four minutes of the late start. The train was two minutes overtime at Kettering and then ran the 72 miles to London in 76 minutes (75 net), arriving just under five minutes late. The pair averaged 76mph on the descent from Sharnbrook and held 52 on the long 1 in 200 to Leagrave summit. Signal checks in the Hendon area prevented more significant time recovery.

All ten were taken into BR ownership in 1948, three being condemned shortly after nationalisation (492 in 1948, 483 and 490 in 1949). The others were subsequently withdrawn as follows: 40488 from Lancaster in 1950, 40484 from Skipton (1953), 40485 from Leicester and 40486 from Saltley (1957), 40489 from Gloucester (August 1960), 40491 from Leeds (September 1960) and 40487 from Nottingham in January 1961.

Leicester's 490, the former 1674, rebuilt with an 'H' boiler in September 1906, runs into Blaby station with the 2.05pm Leicester – Hereford stopping train, 20 April 1911.
(MLS Collection)

No 491, rebuilt in 1913, was one of thirteen class 2 4-4-0s converted to oil-burning in April 1921 as a result of the miners' strike, and reverted to coal-burning in July. It is heading a Nottingham-St Pancras express past Ratcliffe Junction, May 1921. (A.C. Gilbert/MLS Collection)

No 488, rebuilt from the former 1672, with an inspection saloon crossing the Lune River at Tebay, c1948. It was withdrawn from Lancaster shed in December 1950. (MLS Collection)

No 40491 runs into Chinley station with the 12.50pm Sheffield – Manchester, 30 July 1955. (A.C. Gilbert/MLS Collection)

No 40487, the rebuilt 1671, at Chinley North Junction with a Manchester-Sheffield stopping train including an LNER Gresley and GW Hawksworth coach in the formation, 21 August 1959. No 40487 was withdrawn from Nottingham shed in January 1961. (MLS Collection)

Gloucester's 40489 pilots Jubilee 45626 *Seychelles* into Gloucester Central with a Newcastle-Cardiff express, 7 August 1959. No 40489 was withdrawn from that shed in the Spring of 1960. (MLS Collection)

No 209 of the '150' series built in December 1897, rebuilt with an 'H' boiler in August 1906, renumbered 502 in 1907 and rebuilt as a superheated '483' class in May 1912, at Manchester Trafford Park, c1900. (MLS Collection)

The 150 class (LMS 493-502)
Design & construction

Another order for 7ft wheeled 4-4-0s was made in November 1896, ten to replace withdrawn locomotives, and therefore assigned to the revenue account, were delivered between September and December 1897. They therefore were numbered in gaps in the numerical system rather than being allotted a new series number – 150, 153-155 and 204-209. They were equipped with the type 'D' boiler pressed at 160lbs psi and had 19in x 26in cylinders and piston valves. Most were later lined up to 18½in, only 205 and 207 retaining the 19in cylinders. All had 3,250 gallon tenders.

No 498, rebuilt from the former 205 in March 1913, at Liverpool Brunswick, 21 April 1935. (H.N. James/MLS Collection)

No 40495 at Rhyl, c1956. It was shedded there throughout the 1950s until transferred to Bourneville in March 1957, being withdrawn just four months later. (MLS Collection)

They were all rebuilt with 'H' boilers between August 1906 and October 1907, renumbered 493-502 in 1907 and were all replaced with the '483' superheated design 'rebuild' in 1912 and 1913. In fact, although the new class was designated the '483' class, it was 494 of the '150' class that appears to have been the first completed rebuild in February 1912. All were taken over by BR in 1948 and were withdrawn between December 1948 (494) and February 1961 (40502).

Operations
The initial allocation of these ten locomotives was:

Leicester:	150, 153-155
Saltley:	204
Manchester:	205-209

By 1902, the allocation was similar, the five Manchester engines being based at Trafford Park. After the reboilering, 150 moved to Kentish Town and the other three Leicester engines, 153-155, to Nottingham. No 204 joined 205-208 at Nottingham and 209 was at Bristol.

All ten in rebuilt '483' form were received into BR ownership, although 494 was withdrawn in

No 495 and Compound 1020 at Bugsworth Junction (renamed Buxworth in 1930) on the Manchester-Derby line with an express, c1920. (MLS Collection)

No 494 departs from Colwyn Bay with an excursion for Llandudno, c1935. No 494 was one of the earliest to be rebuilt as a superheated '483' class in February 1912 and was also one of the first to be withdrawn in December 1948. (MLS Collection)

No 499 with a Nottingham – Blackpool special at Chapel-en-le-Frith, 30 March 1939. No 499 was withdrawn from Rowsley shed in 1952. (MLS Collection)

1948 and 496 and 500 in 1949. The other seven were withdrawn as follows: 40498 (1950), 40497 from Spital Bridge (1951), 40499 from Rowsley (1952), 40495 from Bourneville in 1957 (although it had been shedded at Rhyl for most of the decade), 40493 from Nottingham (1959), 40501 from Bristol (1960) and 40502 from Nottingham in February 1961.

The 2421 class (LMS 503-522)
Design & construction

Traffic continued to grow rapidly at the turn of the century and in July 1897, Johnson once more required the Locomotive Committee to order more locomotives – 100 in total, of which 20 were to be passenger 4-4-0s. These went out to tender and 2421-2440 were delivered between October and December 1899 by the Sharp, Stewart Company, allegedly because they offered the earliest delivery date! Although this series was often combined with the ten 1897 4-4-0s and labelled the '150' class, in fact once more there were detailed differences - 18½in cylinders and an increase in boiler pressure to 170lbs psi, still basically the 'D' boiler.

They were all rebuilt with 'H' boilers between December 1906 and August 1908, renumbered 503-522 in 1907 and 'replaced' or rebuilt by '483' class superheated boiler engines between March 1912 and May 1913. As the earliest conversions were made to the most recently built class 2 4-4-0s, one can only assume that the intention had been to retain more of the original engines, such as frames, than the subsequent conversion or replacement of the earlier classes.

No 508, formerly 2426 built in November 1899, one of the first Johnson 4-4-0s to be completely rebuilt in 1912 with superheated boiler and new frames to the '483' design, seen at Weston-Super-Mare after arrival with an excursion from the West Midlands, c1923. (MLS Collection)

No 40518 at Sheffield Millhouses, renumbered after a light Works repair but still with LMS on tender, 1948. No 40518 would be transferred to Leeds in 1953 and be withdrawn in 1956. (R.K. Evans/MLS Collection)

The twenty 2421 series in rebuilt form were all taken into BR ownership after 1948 and were withdrawn between 1949 and 1961, the last survivors being 40504 (ex 2422) and 40511 (ex 2429).

Operations

The entire class of twenty locomotives was delivered to Nottingham depot apart from the last five, 2436-2440, which went to Leeds.

They were all still in the same shed locations in 1902, but after rebuilding with the 'H' boiler, their allocation was as follows:

Nottingham:	503-505, 508-511, 513, 514
Leeds:	515-522
Bristol:	506, 507
Kettering:	512

No 2425 (renumbered 507 in 1907) at Great Rocks Dale with a St Pancras-Manchester express, 1901. (D.F. Tee/MLS Collection)

No 2424 of Nottingham pilots a class 3 'Belpaire' on a down express south of Leicester, c1903. (MLS Collection)

No 503 (formerly 2421) running as an oil-burner during the 1921 miners' strike, at Cheadle Heath. It was converted in April and reverted to coal-burning by July.
(MLS Collection)

No 521 at Barnt Green on the Bristol-Birmingham route, 16 July 1921. (W.L. Good/MLS Collection)

Miles	Location	Cheltenham-Birmingham New Street 521 (superheated) (ex 2439) 225/245 tons 7.20pm Bristol Mail Times	Speeds	Gradients
0	Cheltenham	00.00		
3.8	Cleeve	05.45		
7.2	Ashchurch	09.00	68	1/295 F
9.4	Bredon	11.05	60	1/301 R
13.2	Defford	14.40	67	L
18	Abbots Wood Jcn	19.40	51	1/301 R
20.4	Spetchley	22.30		
24.6	Dunhampstead	26.40	62	L
31.3	Bromsgrove	34.25		
0		00.00 Attach 'Big Bertha' 0-10-0		
2.2	Blackwell	07.20	18 (ave)	1/37¾ R
3.6	Barnt Green	10.00		
8.7	King's Norton	15.55		1/301 F
10.9	Selly Oak	18.40		
14.2	Birmingham	23.50		

No 508 assists Stanier 'Black 5' 5283, on the Lickey Incline, banked by a Jinty 0-6-0T, 1941. (MLS Collection)

No 509 pilots 44839 on the southbound *Pines Express* at Chilcompton, 3 October 1949.
(J.D. Darby/MLS Collection)

All the former 2421 class lasted until the nationalised era, but four (506, 510, 512, and 517) were withdrawn before acquiring their BR number. The others were based at Nottingham, Hasland, Stafford, Nuneaton, Derby, Toton, Leeds, Burton on Trent, two on the S&D at Bath Green and three at Royston, although one of the latter (40520) had been at Rowsley for most of the 1950s. The last survivors were withdrawn from the following depots: 40513 from Derby (1959), 40504 from Nottingham (January 1961) and 40511 from Toton (January 1961) although it was based at Saltley throughout the 1950s.

The 60 class (LMS 523-562)
Design & construction

The last class of Johnson class 2 4-4-0s was initially ordered from the revenue account in June 1898. The locomotives were built at Derby and although authorised after the Sharp, Stewart 2421 series, because of the delayed delivery of the latter, the first of the new class '60' engines were put into traffic first in June 1898. Ten were built in the first batch, numbered 60-66, 93, 138 and 139 in that order, the last completed in July. A second batch of ten, numbered 67-69, 151, 152 and 165-169, were built between July and October 1899, the numbers replacing those of withdrawn Kirtley 2-4-0s. Five more were constructed in 1901 and numbered 2636-2640, and then another five from the revenue account and numbered 805-809. The final order for engines of the class were built by Neilson Reid at a cost of £3,645 each and were numbered 2591-2600, entering service between May and June 1901. The key dimensions for this last Johnson class 2 4-4-0 class were:

Cylinders	19in x 26in (most were later lined up to 18½in)
Coupled wheel diameter	7ft 0½in
Bogie wheel diameter	3ft 6½in
Piston valves	8in diameter
'E' type boiler pressure	170lbs psi
Heating surface	1,205sq ft
Grate area	19.6sq ft
Axleload	16 tons 19 cwt
Weight – Engine	47 tons 4 cwt
– Tender	38 tons
– Total	85 tons 4 cwt
Water capacity	3,500 gallons
Coal capacity	4 tons
Tractive effort	14,440lbs

Only 805-809 and 2593-2597 plus 2600 retained the larger 19in diameter cylinders. After rebuilding with 'H' boilers between 1906 and 1908, the engine weight increased to 51 tons. The engines were renumbered 523-562 in 1907. They were replaced/rebuilt as the '483' superheated class between February 1913 (534) and March 1915 (523). A number were equipped as oil burners during the miners' strikes in 1921 and 1926.

All were taken into BR ownership in January 1948. The first of this series to be withdrawn was 545 in December 1948, and the last survivor was 40537, withdrawn in September 1962.

No 169, built in 1899, at Bristol Temple Meads, c1900. (Locomotive and General/MLS Collection)

One of the '60' class, No 63, built in 1898, in Works grey for the official photograph. (Railway Photographs/MLS Collection)

No 65, built in 1898, at St Pancras, c1900. It was rebuilt with an 'H' boiler in October 1906 and replaced with a superheated boiler engine in May 1913. (E.M. Johnson Collection)

No 550, formerly 2638, as rebuilt with superheater boiler in January 1915. (E.M. Johnson Collection)

No 559, rebuilt as a '483' superheated engine in December 1914, converted to oil-burning during the 1926 'General Strike', at Kentish Town, 15 July 1926. (H.C. Casserley/MLS Collection)

No 40562, the last of the Johnson 4-4-0s built for the Midland Railway and numbered 2600, rebuilt in 1913 with superheated boiler, at Skipton in early BR mixed traffic livery, 13 May 1949. (MLS Collection)

Derby's 40538 at its home depot alongside Stanier Black 5 44945. Rebuilt in January 1914, it was one of the last survivors of the Midland Railway 2Ps, withdrawn in May 1959. (MLS Collection)

Operations

60-66 were allocated at first to Leicester and 93, 138 and 139 to Bristol. Nos 67-69 were also allocated to Leicester but the remainder of the second batch went to Derby. By the completion of the delivery of the forty engines the allocation was:

Leicester:	60-69, 2591, 2594, 2595, 2600
Derby:	151, 152, 165-169
Bristol:	93, 138, 139
Carlisle:	805-809
Leeds:	2636-2640
Kentish Town:	2592, 2593, 2596-2599.

By 1902, little had changed – Leicester's 2591 had moved to Derby but all Kentish Town's allocation went to Leicester. After reboilering, the 1908 allocation was:

Leicester:	60-63, 65-68, 2592-2599
Derby:	64, 69, 2591
Bristol:	93, 138, 139, 165-167
Nottingham:	151, 152
Saltley:	168, 169
Carlisle:	805-809, 2636-2638
Lancaster:	2639, 2640
Kentish Town:	2600

Leicester-St Pancras, 1907 / 8

		526 (ex 63) 250 tons		525 (ex 62) 185 tons			
Miles	Location	Times	Speeds	Times	Speeds		Gradients
0	Leicester	00.00		00.00	T		
3.7	Wigston	07.09		06.30			
7.3	Great Glen	12.16	pws	-	pws		1/199 R
10.2	Kibworth	15.51		14.30			
16	Market Harborough	21.44		21.40			1/238 F
20.8	Desborough	27.51	40	27.10	45		1/132 R
27	Kettering	33.13	80	32.50	72	2¾ L	1/118 F
34	Wellingborough	38.42		38.50	70		
36.3	Irchester	40.50		-			
39.3	Sharnbrook summit	-	40½	45.20	45		1/120 R
42.3	Sharnbrook	47.41	76½	-			1/119 F
49.2	Bedford	53.57	sigs 38*	55.10	75	3¼ L	
57.3	Ampthill	64.42	40	64.10	50		1/200 R
61.8	Harlington	70.46		-			1/200 R
66.3	Leagrave	77.29	38½	75.35	45		1/200 R
68.8	Luton	80.11		78.20		5¼ L	
74.4	Harpenden	85.29		-	easy		
79.2	St Albans	89.47		89.10			
83.8	Radlett	93.27	78	93.10	70		1/176 F
86.5	Elstree	96.02	62	96.10	45		1/200 R
92	Hendon	101.02		101.20			
94	Cricklewood	102.39	sigs	-			
97.5	Kentish Town	107.44	sigs	106.25			
99	St Pancras	<u>112.20</u>	<u>7¼ L</u>	<u>108.30</u>		<u>3½ L</u>	
		(106¾ net)		(106½ net)			

R.E. Charlewood and Cecil J. Allen, describing the runs in their articles in the *Railway Magazine,* found the 105 minute schedule too tight for a class 2 when overloaded. The former 62 and 63 had by this time been rebuilt with an 'H' boiler in 1906. No 562 (the former 2600) had also been reboilered (in 1907) but the schedule of the train logged below had been extended to 110 minutes.

St Pancras-Leicester
562
193/205 tons

Miles	Location	Times	Speeds		Gradients
0	St Pancras	00.00		T	
1.5	Kentish Town	05.08		(fog) 1 L	1/178 R
6.9	Hendon	13.45		3 ¾ L	1/200 F
12.4	Elstree	20.24	47		1/176 R
15.2	Radlett	23.13	66½		1/200 F
19.9	St Albans	27.55	46	3 L	1/176 R
24.6	Harpenden	33.45			
30.2	Luton	39.22		3¼ L	
32.8	Leagrave	41.53			
37.3	Harlington	45.57	82		1/200 F
41.8	Ampthill	49.24	76		1/200 F
49.8	Bedford	55.26	84	1½ L	
56.7	Sharnbrook	61.43			1/119 R
59.7	MP 59 ¾	66.04	37½	1 L	1/119 R
65	Wellingborough	70.55	75*	1 L	1/120 F
72	Kettering	78.00		1 L	
78.5	Desborough North	86.10	45		1/136 R
82.9	Mkt Harborough	90.39	72*	½ L	1/132 F
86.3	East Langton	94.44	59		
89.7	Kibworth North	98.35	46		1/130 R
95.4	Wigston	104.05	69*		1/199 F
99.1	Leicester	108.48		1¼ E	

Around 1911, No 555 (the former 2593), rebuilt in January 1907 with an 'H' boiler, was working a Nottingham-London train via Oakham. Although the load was only 175 tons, it was provided with the assistance of Johnson 'Single' No 605 on the adverse grades as far as Melton Mowbray where the pilot engine was detached, having covered the 18¼ miles in five seconds less than the tight schedule allowed. After that, 555 was on its own and took things fairly easily to Kettering (38 minutes 35 seconds for the 33¼ miles), gaining just a minute of the time lost through the time taken to detach 605 at Melton. The crew changed at Kettering and considerably more energy was shown, 555 running up to

London in 76 minutes (74 net), the highlights being 70½mph reached before Wellingborough, a minimum of 45½mph up to Sharnbrook summit and 79 on the descent through Bedford, a p-way slack before Ampthill and recovery to 48½mph up the long 1 in 200 to Leagrave, 75 after St Albans down to Radlett and again at Hendon. Arrival in St Pancras was 3 minutes late. No 555 was further rebuilt with a G7 Belpaire boiler in 1913.

A *Railway Magazine* article in February 1914 records the run of a recently rebuilt superheated class 60, 549, renewed in June 1913, between Leeds and Rotherham on an overloaded 2.50pm Leeds-St Pancras express. The 33¾ miles is difficult with numerous colliery subsidence slowings and was scheduled to be completed in 45 minutes. No 549 accomplished the journey in 44 minutes 25 seconds, with 300 tons, (43½ net) with a maximum speed of 65mph.

Most had been rebuilt as '483' superheated engines by this time, but in the summer of 1914 Bristol had a couple of unrebuilt engines with 'H' boilers – 60 and 61 – and Leicester had the majority – 138, 167, 806, 2638 and Kentish Town had just one unrebuilt engine, 2597.

Although rarely used for any work other than piloting 'Black 5s' or 'Jubilees' or hauling light stopping services after nationalisation which meant few records of runs of interest exist, 40551 was rostered to a 5-coach Institute of Locomotive Engineers special train in the early 1950s and showed a surprising turn of speed.

'H' boilered 550 with tall chimney and dome on a down express emerging from Elstree Tunnel, c1912. (H. Gordon Tidey/MLS Collection)

Kentish Town's 554 rebuilt with an 'H' boiler in September 1907 and 'replaced' with superheater boiler in December 1913, on an up express south of Bedford, c1912. (F. Moore/MLS Collection)

Bristol's 523 rebuilt with an 'H' boiler in August 1906 and replaced by a superheated boilered engine in March 1915, with the 3.50pm Bath Green Park-Bournemouth train, 26 July 1913. (E.M. Johnson Collection)

		Kettering-St Pancras 40551 5 chs, 166/175 tons		
Miles	Location	Times	Speeds	Gradients
0	Kettering	00.00	T	
3.8	Finedon	05.53		
7	Wellingborough	09.01	57½ / pws 20*	
9.3	Irchester	11.54		
12.3	Sharnbrook summit	17.20	37	1/120 R
15.4	Sharnbrook	20.32	7½	
19	Oakley	23.29	68	
22.1	Bedford North Jcn	26.05	71	
24.7	Elstree	28.29	65	
30.2	Ampthill	34.15	52	1/200 R
31.8	Flitwick	35.54	60/pws 20*	
34.7	Harlington	40.55		
38	MP 34	45.09	51	1/200 R
39.2	Leagrave	46.30	66	
41.8	Luton	48.58	61	
44.7	Chiltern Green	51.52	70½	1/176 F
47.4	Harpenden	54.06	64½/69	1/200 R, 1/176 F
52.1	St Albans	58.21	64½*/69	1/176 F
56.8	Radlett	62.17	75	1/176 F
59.6	Elstree	65.00	55	1/200 R
62.7	Mill Hill	67.57	68	1/176 F
65.1	Hendon	69.59	72½	
70.5	Kentish Town	75.34		
72	St Pancras	77.22		3½ E

Seven of the class were withdrawn in the early years of nationalisation before renumbering, and six lasted into the 1960s – 40552 withdrawn from Leeds Holbeck (July 1960), 40543 from Leicester and 40548 from Kentish Town (January 1961), 40557 from Nottingham (March 1961), 40540 from Gloucester (February 1962) and the final survivor, 40537, from Bristol Barrow Hill in September 1962. In the 1950s the main distribution had been at the following locations (just a single one at each depot unless otherwise specified):

Preston
Longsight
Crewe
Chester
Llandudno Junction
Nuneaton
Northampton
Hasland (2)

Sheffield Millhouses
Leeds Holbeck
Manningham (Bradford)
Burton on Trent (2)
Walsall
Bourneville
Spital Bridge (2)
Bristol Barrow Road

Gloucester (4)
Nottingham (6)
Derby
Leicester (2)
Bedford
Kentish Town (3)
Templecombe

Gloucester's 40523, rebuilt from the first of the class 60s built in 1898, pauses amid its duties at Cheltenham Lansdown station, May 1950. It would be withdrawn in October 1952. (W. Potter/MLS Collection)

Nuneaton's 40528, formerly No 65, built in 1898, rebuilt as a superheated 2P in May 1913, at Crewe with a parcels train coming off the Manchester line, 23 June 1951. No 40528 was withdrawn in December 1952. (T. Lewis/MLS Collection)

No 40537 and BR Standard 5 73049, both allocated to Bath Green Park shed, arriving at Bath from Bournemouth on the *Pines Express*, 18 March 1959. No 40537 was the last survivor of the Midland class 2 4-4-0s, withdrawn in September 1962. (David Maidment)

The Somerset & Dorset Railway 4-4-0s

The financially stretched independent Somerset & Dorset Railway bowed to the inevitable in 1875 and leased the line to the Midland and London & South Western railways from 1 November. The Midland agreed to provide the locomotives, the South Western the rolling stock and civil and signal infrastructure. After fifteen years of utilising 0-4-4 tank engines for its main services, provided under Johnson at Derby, the local superintendent, Whitaker, requested four passenger engines, querying the value of more tank engines. Johnson concurred and recommended four 4-4-0s similar to the 1808 class currently under production for the Midland at Derby, but with 5ft 9in coupled wheels in place of 6ft 6in.

Ordered in December 1889 at a cost of £1,950 each, numbers 15-18 were not delivered until May 1891. Key dimensions were:

Cylinders (2 inside)	18in x 24in
Coupled wheel diameter	5ft 9in
Bogie wheel diameter	3ft 0in
Boiler pressure	150lbs psi
Heating surface	1,202sq ft
Grate area	16sq ft
Axleload	15 tons 2 cwt
Weight – Engine	39 tons
– Tender	29 tons 18 cwt
– Total	68 tons 18 cwt
Water capacity	2,200 gallons
Coal capacity	3 tons

In 1895, four more were ordered, two in February and two in October, with 67 and 68 being delivered in January 1896 and 14 and 45 in February 1897.

As loads increased, these eight modestly powered 4-4-0s struggled on a route that was vastly more difficult than the Midland main line on which their sisters performed satisfactorily. Although Derby was being pressed for something more powerful, when reboilering was necessary, similar sized boilers were provided albeit at 160lbs psi. Eventually, Derby was persuaded to rebuild them with the larger 'H' boilers fitted between 1907 and 1911. The grate area was increased to 21.1sq ft, heating surface varied between 1,282 and 1,353sq ft, the axleload increased to 16 tons 6 cwt and engine weight to 46 tons 8 cwt. A 2,600 gallon tender was provided weighing 32 tons 18 cwt.

No 18, Johnson 5ft 9in 4-4-0 of 1891, similar in other dimensions to the Midland 1808 class, S&D class A, at Bournemouth West on an express train for Bath, c1895. Note the headlamps acting as the LSWR route indicator rather than denoting express or stopping train headcode. (MLS Collection)

No 45, Johnson slim boilered 5ft 9in 4-4-0 built in 1897, struggling through the snow near Chilcompton in the winter of 1909, months before reboilering with the Derby 'H' boiler. (Locomotive & General/MLS Collection)

No 68, built in 1896 with 5ft 9in coupled wheels but other similar dimensions to the Midland 2183 class, rebuilt in 1908 with 'H' larger boiler and cab, leaving Radstock station, c1912. It was withdrawn in 1921 and replaced by a superheated 4-4-0 with the same number. (Locomotive & General/MLS Collection)

Johnson 1896 built 4-4-0 No 67, rebuilt with 'H' boiler in 1907, at Bournemouth West, c1910. It was withdrawn in 1921 and replaced by a superheated boiler 4-4-0 with the same number. (MLS Collection)

No 45, rebuilt with Derby 'H' boiler in 1909, at Bournemouth West, c1923. In 1925, it was rebuilt with a Belpaire boiler, renumbered 18 in 1928 and LMS 303 in 1930. (F. Moore/MLS Collection)

No 18, the former 45, was rebuilt in 1926 with Deeley G7 boiler and Belpaire firebox, seen here at Derby immediately after conversion. To add to the confusion, it was renumbered 15 in August 1928. (F. Moore/ MLS Collection)

LMS 301, the former S&D 18, then 15, in LMS livery in front of Fox Walker saddle tank 1504 (ex S&D No 5 of 1875), at Highbridge, 5 July 1930. (H.C. Casserley/ MLS Collection)

LMS 303, the former S&D 45, then 18, rebuilt in 1925 with a Belpaire boiler, at Bath shed, c1931. (F. Moore/MLS Collection)

Nos 67 and 68 were withdrawn in 1921 and replaced by two superheated 4-4-0s with identical numbers, but the other six continued on less demanding duties. No 45 was rebuilt with a G7 Belpaire boiler in 1926. No 17 was treated likewise in 1927.

In November 1903, Derby delivered three larger 4-4-0s ordered in August 1902, similar to the contemporary Midland 2581 series with 6ft coupled wheels. They were numbered 69-71 and their key dimensions were:

Cylinders (2 inside)	18in x 26in
Coupled wheel diameter	6ft 0in
Bogie wheel diameter	3ft 1in
Boiler pressure	175lbs psi
Heating area	1,420sq ft
Grate area	21.1sq ft
Axleload	16 tons 9 cwt
Weight – Engine	46 tons 4 cwt
– Tender	35 tons 2 cwt
– Total	81 tons 6 cwt
Water capacity	2,950 gallons
Coal capacity	3 tons

For some reason these three impressive looking engines were not as successful as the reboilered earlier engines and 70 and 71 were withdrawn in 1914 and replaced by superheated 4-4-0s with the same numbers. The decision to replace 69 also was delayed by the war situation, but it was withdrawn in 1921 and replaced then similarly by a superheated 4-4-0. How much of the withdrawn engines was incorporated in the new locomotives is unknown.

Two more 4-4-0s were ordered from Derby in May 1907 and 77 and 78 were delivered in 1908. This pair had Derby 'H' boilers, Deeley cab and were similar otherwise to the 1903 engines apart from:

Heating surface	1,347sq ft
Axleload	16 tons 12 cwt
Weight – Engine	47 tons 8 cwt
– Tender	36 tons 13 cwt

These engines were more powerful although they and the 1903 engines had a heavier coal consumption than the reboilered 1891 and 1896 engines.

No 71, a Johnson large boilered 4-4-0 with 6ft coupled wheels, built at Derby in 1903, was withdrawn in 1914 and replaced by a superheated 4-4-0 with the same number, later 40, and LMS 323, c1912. (Real Photographs/MLS Collection)

Deeley 1908 4-4-0, No 77, built with 6ft coupled wheels and 'H' boiler and Belpaire firebox, still resplendent in S&D blue livery, at Bath shed, 24 May 1929. It was withdrawn as LMS 320 in 1931. (H.C. Casserley/MLS Collection)

The Midland Class 2 • 115

LMS 321, Deeley 1908 4-4-0, formerly S&D 78, at Bath depot, c1932. LMS 2P 332, rebuilt from Midland Railway 1566 of 1882, is behind in the covered shed. (Locomotive & General/MLS Collection)

A second view of LMS 321, with Midland standard tender, on shed after absorption by the LMS, c1935. (MLS Collection)

The replacements for the two S&D 4-4-0s, 70 and 71, withdrawn in 1914 were superheated 4-4-0s similar to the Midland Railway 483 class, at a cost of over £2,800. No 69 was similarly dealt with in 1921 along with 67 and 68, the costs having escalated post-war to around £8,000 each. The dimensions of the basically new engines were:

Cylinders (2 inside)	20½in x 26in
Coupled wheel diameter	7ft 0½in
Bogie wheel diameter	3ft 6½in
Boiler pressure	160lbs psi
Heating surface	1,170sq ft
Superheating surface	313sq ft
Grate area	21.1sq ft
Axleload	17½ tons
Weight – Engine	53 tons 7 cwt
– Tender	37 tons
– Total	90 tons 7 cwt
Water capacity	3,250 gallons
Coal capacity	4 tons

The new superheated engines were more economical than the 1903 and 1908 Derby engines, reducing coal consumption per mile by around 25 per cent, though, like all the 4-4-0s, they were really underpowered for the S&D gradients so that piloting of trains, especially in the summer months, was rife. Few accounts seem to have been made of their running or that of the earlier engines. Some details from 1914 show the various earlier 4-4-0s surmounting the 1 in 50 to Masbury summit with around 150-160 tons at no more than 15mph, while a superheated 4-4-0 just about made 20mph. Two of the new superheated 4-4-0s, 67 and 68, received oil burning equipment in 1921 and worked successfully for the duration of the national coal strike albeit the oil pollution in the area was significant.

Tests took place in 1925 to find more powerful locomotives for the route. Midland 4P 995 was tried, but it used more fuel than the superheated 2Ps. Then, in 1927, Horwich 'Crab' 13064 was tested and was very popular with the local crews but, surprisingly, the apparently successful trial was not followed up. Then, in 1928, three new Fowler 2P 4-4-0s (575, 576 and 580) were allocated to the S&D and renumbered 44-46 (see Chapter 7, page 266).

No 69 was a Johnson large boilered 4-4-0 similar to the Midland 2581 series with 6ft coupled wheels, built at Derby in 1903, withdrawn in 1921 and replaced by this Belpaire boilered superheated 4-4-0 with the same number at a cost of £7,625. It was renumbered 43 in 1928 and LMS 326. (F. Moore/MLS Collection)

The Midland Class 2 • 117

The superheater 4-4-0 that replaced 1903 built 71 in 1914, here in S&D blue livery in the early 1920s. It was renumbered by the LMS as 323 in 1930. (F. Moore/ MLS Collection)

The *Pines Express* at Bath Junction behind superheated 4-4-0 No 41, built as No 67 in 1921 to replace the 1903 No 67, c1923. Through services had Midland Railway rolling stock whilst the services restricted to the S&D were provided with L&SWR coaches. (Locomotive & General/ MLS Collection)

LMS 323 (ex S&D No 71), built in 1903 and rebuilt with G7 superheated boiler for the S&D as No 71 in 1914, at Manchester Central, May 1937. (W. Potter/MLS Collection)

LMS 323 (ex S&D 71) at Sheffield Midland with a local stopping train, c1935. (Photomatic/MLS Collection)

The Midland Class 2 • 119

BR (LMR) 40323 (ex S&D 71) at Derby, renumbered but still retaining LMS inscribed on the tender, 10 July 1948. (W.L. Good/MLS Collection).

No 40326 in BR mixed traffic lined black livery at Buxton shed, c1952. (W. Potter/MLS Collection)

An M&GN 4-4-0, No 14, designed by S.W. Johnson at Derby similar to the Midland 1808 class and the Somerset & Dorset class 'A', constructed by Sharp, Stewart & Co. in 1894, taken at Cromer, c1905. No 14 remained unrebuilt and was withdrawn by the LNER in February 1937. (Locomotive Publishing Co./MLS Collection)

Once the S&D working had been taken over by the LMS in 1930, they returned to the LMS fleet renumbered 633-635 and in 1933 one of them, 633, received the experimental Dabeg feed-water treatment. An increasing number of LMS 2Ps such as 30563/4/8, 40601, 40698, 40699 and 40700 were allocated to the S&D, and some of the former S&D locomotives found themselves transferred elsewhere on the LMS system.

The small 4-4-0s built in the 1890s had all been withdrawn by 1932 (No 45, renumbered 18 was the last), the 1908 constructed 77 went in 1931 and 78 in 1938. After the replacement in 1914 and 1921 of the 1903 4-4-0s, their superheated replacements remained until 1936, when 322-324 moved to Sheffield Millhouses and 326 and 326 to Saltley. The five engines became BR property in 1948, renumbered 40322-40326, lasting to between 1951 (40325) and September 1956 (40326). They ran more than a million miles in traffic, 40322 recording 1,012,139 miles and 40323, 1,110,365.

The Midland & Great Northern Railway 4-4-0s

The Midland and Great Northern railways assumed the working of trains on the Eastern & Midlands Railway west of King's Lynn from 1889 and the complete system of Peterborough to Melton Constable, Cromer and Yarmouth from 1893, forming the M&GN Joint Committee to run the railway. In 1894, the Committee procured from the Sharp, Stewart Company some Johnson inside cylinder 4-4-0s of its own to dispense with the use of Midland and Great Northern locomotives – engines that were similar to the 1808 class being built at the time for both the Midland and Somerset & Dorset railways. Twenty-six of these class 'C' locomotives were built in 1894, numbered 1-7, 11-14, 17, 18, 36-39 and 42-50. Seven more, 51-57, were delivered in 1896 and another seven, 74-80, the latter built by the Beyer, Peacock Company, in 1899. Their dimensions were:

Cylinders (2 inside)	18½in x 26in
Coupled wheel diameter	6ft 6½in
Bogie wheel diameter	3ft 3½in
Boiler pressure	160lbs psi
Heating surface	1,078sq ft
Grate area	17.5sq ft
Axleload	16 tons
Weight – Engine	42 tons 18 cwt
– Tender	33 tons 11 cwt
– Total	76 tons 9 cwt
Water capacity	2,950 tons
Coal capacity	3 tons
Tractive effort	15,416lbs

The only differences from the Midland 1808s were the half-inch larger diameter cylinders and deeper frames. The Beyer, Peacock engines to the same design were the blueprint for the Midland Railway's 2581 class built by the same company in the same year.

Twenty-three of the locomotives remained unaltered apart from boiler replacements of similar dimensions although they received Midland type smokebox doors and detailed alterations to cab and chimneys. They were taken into LNER stock in October 1936 and classified D52.

Seven of the engines – 2, 6, 36, 44, 49, 60 and 77 – were rebuilt with Midland G6 boilers between 1929 and 1931 and were classified by the LNER as D53s. As early as 1908, two (39 and 55) were reboilered with the 'H' boiler and between 1910 and 1925 these two and eight more – 45, 46, and 51-57 – were rebuilt with Belpaire G7 boilers and in 1936 were classified as D54s. The D53s weighed 30 cwt heavier, the D54s had boiler pressure of 175lbs psi, heating surface of 1,384sq ft and a larger grate area of

Johnson 4-4-0 No 3, built in 1894 as No 14 above, but reboilered with a small second-hand Derby boiler with extended smokebox, Midland style smokebox door and tall narrow chimney in 1932, withdrawn in June 1937. The LNER classified these 1894 locomotives as D52. (Photomatic/MLS Collection)

Johnson 4-4-0 M&GN No 1, built in 1894 and rebuilt with a Midland/LMS small boiler in 1932, on a stopping passenger service on the M&GN system, c1935. It was withdrawn in November 1937. (Locomotive & General/MLS Collection)

21sq ft. Tractive effort was thereby increased to 16,862lbs. The D54s were mainly on the Peterborough-Cromer and Leicester-Yarmouth and Lowestoft through services, the D52s and D53s working the stopping passenger services, fish and goods trains.

The LNER reviewed the stock they had inherited and withdrew thirteen D52s, one D53 and three D54s between November 1936 and November 1937. They renumbered all the M&GN engines on the 'duplicate' list with the '0' prefix. The last D52 was 038 withdrawn in September 1943, the last D54s were 055 and 056 withdrawn in November 1943 and the last of all, D53s 050 and 077, withdrawn in January 1945 before they received their allocated LNER numbers of 2052 and 2054. (06 allocated 2053 was withdrawn in 1944.)

Johnson 4-4-0 No 77 rebuilt with Belpaire G6 boiler in December 1930, a D53 renumbered 077 on the duplicate list when the LNER took responsibility for M&GN locomotives in October 1936, seen here approaching Nottingham a few months earlier, 19 May 1936. It was withdrawn in January 1945. (G.A. Barlow/MLS Collection)

M&GN No 53, rebuilt in 1910 with Deeley G7 boiler, at Thurmaston with a Birmingham-Yarmouth train, 4 June 1911. (MLS Collection)

Johnson 1896
4-4-0 M&GN No 54, rebuilt in 1914 with Deeley G7 boiler, extended smokebox and Belpaire firebox, c1930. It was withdrawn as LNER D54 class in October 1939. (Locomotive & General/MLS Collection)

D54 052 with Deeley G7 boiler, after absorption by the LNER in 1936, at Yarmouth Beach shed, 4 September 1938. (MLS Collection)

Chapter 3
THE MIDLAND CLASS 3

Design & construction

The Midland Railway faced a problem of their own making in 1897. They had introduced excellent new bogie carriages, but these clearly weighed significantly heavier (23 per cent more) than previous rolling stock. Asked for a report on the impact of these on motive power requirements, Johnson's assistants intimated somewhat unhelpfully, 'If the pilot mileage is to be reduced and the trains run to time, the weight of the trains must be kept down.' Johnson had already ordered more slim-boilered 4-4-0s, but something more powerful was required and he persuaded the Locomotive Committee to authorise twenty express passenger engines in February 1899.

There was some confusion as to whether the new engines could be accounted for under the Capital or Revenue account, and 2606-2610, the first five of the new engines, were transferred from Capital to Revenue account taking the numbers of five 'Singles' which were renumbered rather than withdrawn. They were then transferred back to the Capital account when the next five, 804-809, were authorised under the Revenue account. Nos 2606-2610 had in fact entered service between September and November 1900 and 804-809 between January and June 1901. The design was based on that of the latest class '60' 4-4-0s, but with 6ft 9in coupled wheels and a larger Belpaire boiler – in fact the new engines were known as 'Belpaires' throughout the company. Their key dimensions were:

Cylinders	19½in x 26in
Coupled wheel diameter	6ft 9in
Bogie wheel diameter	3ft 6½in
Boiler pressure	175lbs psi
Heating surface	1,519sq ft
Grate area	25sq ft
Axleload	18 tons 5 cwt
Weight – Engine	51 tons 17 cwt
– Tender	52 tons 7 cwt
– Total	104 tons 4 cwt
Water capacity	4,000 gallons
Coal capacity	5 tons
Tractive effort	16,477lbs

The next ten engines, built between January and June 1902, were numbered 2781-2790 and allocated to the Capital account and were of a modified design. The main difference was a redesigned boiler, classified G8 (the earlier engine boilers were GX), and were built to the higher pressure of 180lbs psi and slightly increased heating surface of 1,528sq ft. They had a longer wheelbase, smaller diameter bogie wheels of 3ft 3½in and weighed heavier, with an axleload just kept to the required 18½ tons. Three more batches of the class to the 2781 design were ordered between 1901 and 1903, treated as from the Revenue account and delivered as follows:

810-819:	October-December 1902
820-829:	April-September 1903
830-839:	December 1903-February 1904.

Johnson finally ordered a further thirty, numbered 840-869, making eighty of the class in total, and these were delivered between June 1904 and September 1905. The last twenty had a slightly modified G8A boiler with ten fewer tubes and a heating surface of 1,472sq ft. The fleet was renumbered in 1907 in the order of building, 700-779. They were classified as class 3 in 1906 and became LMS class 3P.

The tenders for the last forty locomotives had different capacities. Nos 830-849 had 4,500 gallon tenders and the last twenty, initially 4,100 gallons. However, after

No 2607, the second Johnson class 3 'Belpaire' and 4,000 gallon bogie tender, built in October 1900, in Works grey livery, 1900. It was later renumbered 701 and was withdrawn in 1931. (E.M. Johnson Collection)

No 2788 from the second batch or 'Belpaires' built in April 1902. It has the large bogie tender and smokebox door ring and dart and is in the Johnson lined red livery, and was photographed at Kentish Town, c1902. It was later renumbered 717, was superheated in 1919 and withdrawn in 1935. (Colling Turner/MLS Collection)

No 810 (later 720) from the third batch of 'Belpaires' built in October 1902, seen shortly afterwards at Durranhill, Carlisle, c 1903. (Loco Publishing Co./MLS Collection)

No 863 (later 773) from the final batch of 'Belpaires' built in July 1905. It has a six-wheel tender following the building of water troughs on the company's main lines and a flat smokebox door. It is seen here shortly after construction in 1905. (E.M. Johnson Collection)

No 730 with Deeley smokebox door design and six-wheel tender in the simplified Deeley Midland red livery. It was built in April 1903 and is seen here c1910.
(E.M. Johnson Collection)

No 758 built in 1904, at Kentish Town, c1912, before being rebuilt with a G8A boiler in 1916 and a superheated boiler in 1925. It was one of the last 3Ps to be withdrawn in 1951. It has a six-wheel tender with coal rails.
(MLS Collection)

No 740 at Holbeck with a Midland Compound behind, c1912. It was rebuilt with a G8A boiler in 1915 and superheater in 1922. (MLS Collection)

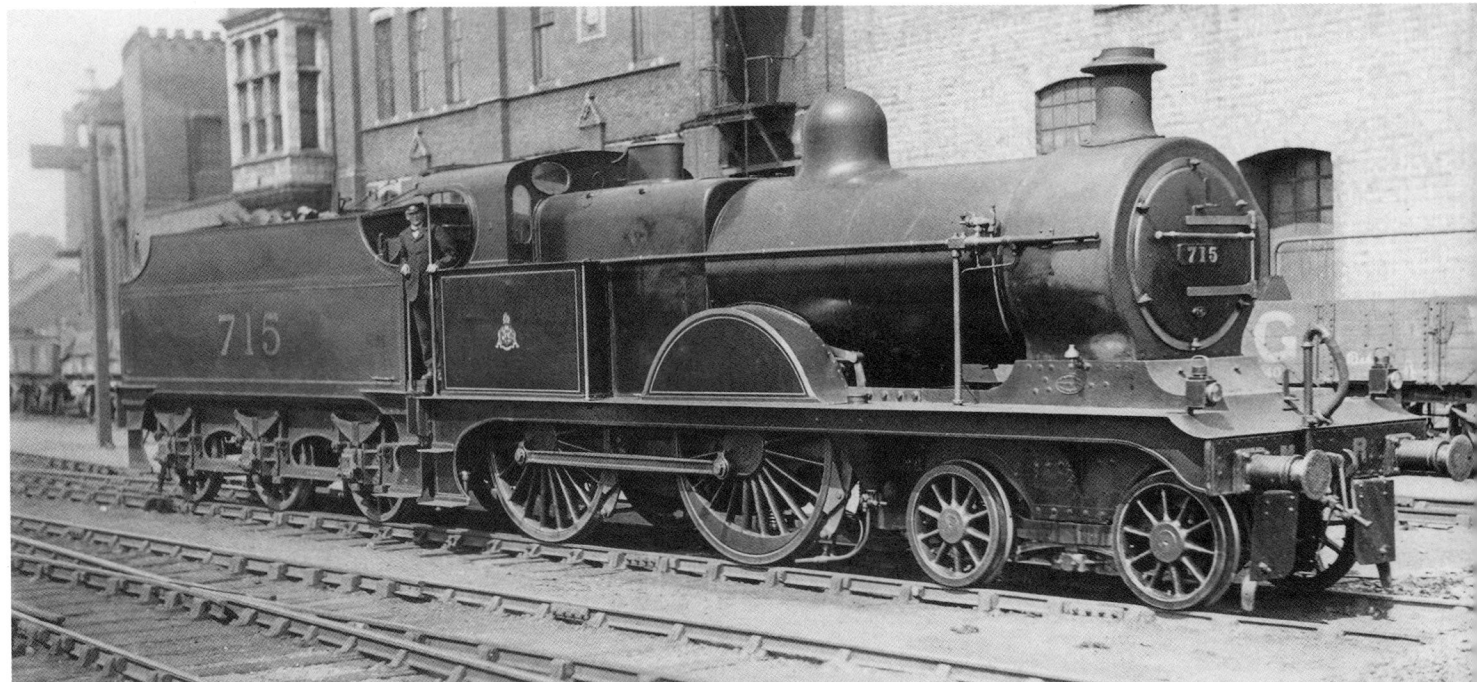

No 715 (ex 2786) with six-wheel rebuilt bogie tender, c1920. It was superheated in June 1921 and withdrawn in January 1948. (MLS Collection)

the routes were equipped with water troughs, six-wheel 3,500 gallon tenders sufficed. Many of the bogie tenders had their water tanks shortened and fitted on the six-wheel frame. All received the standard Midland red livery with the Johnson elaborate lining reverting to the more simplified Deeley style. The red livery was maintained initially by the LMS until 1928 when it became black with red lining. During the Second World War, the survivors were in plain black.

A proposal was made to superheat a locomotive of this class as early as 1906, but it was not implemented. A decision was made later in 1912 to superheat members of the class when their boilers needed replacing, noting in particular the need to provide the first ten with new front ends and wheelbase in line with the rest of the class. Ten new superheated boilers were built between 1912 and 1914, classified G8AS, the first rebuilding being of the prototype, 700 (previously 2606). Seven more of the first ten were rebuilt by 1914, but two were dealt with much later, 706 (1921) and 707 (1925). No 707 had been reboilered in 1910 prior to the superheating decision. The war interrupted the reboilering of the later engines and whilst a few were reboilered with superheating between 1916 and 1919, most were superheated in the 1920s, although withdrawals began before completion. Nos 737, 742, 749, 751, 772, 778 and 779 were withdrawn still saturated in 1925 and 1926.

The coal strikes of 1921 and 1926 caused many Midland engines to be converted to oil-burning. Three engines – 762, 768 and 772 – had been briefly so-equipped during an earlier coal strike in 1912. In 1921 twenty-three engines of the class were oil-burners (six saturated and seventeen superheated). Two 450 gallon oil tanks were mounted in the tender. The strike lasted from April to June. During the longer 'General Strike' in 1926, only one 3P was converted – 765 – as withdrawal of the class had already commenced.

700 and 768 were fitted with feed water heaters in 1913 and 700 especially showed significant savings in coal and water. No. 700 retained the equipment until withdrawn in 1927 and that of 768 was removed around the same time. It seems strange that, in view of the savings made, the equipment was not standardised, although the investment costs involved during and after the First World War made

No 776 was rebuilt with a superheated boiler in July 1920 and was photographed at Derby in April 1921 fitted for oil-burning during the coal miners' strike of that Spring. The two 450 gallon fuel tanks mounted on the tender are very obvious in this official photograph. (MLS Collection)

One of the six saturated 'Belpaires' that was fitted to burn oil in the 1921 strike, 777. It was rebuilt with a superheated boiler in 1924. (E.M. Johnson Collection)

No 768 which was superheated in 1919 and fitted as an oil-burner in 1921. It was one of two of the class fitted with G & F Weir's feedwater heating system in 1914 and the reservoir is seen clearly in this photograph. (E.M. Johnson Collection)

The Midland Class 3 • 131

The prototype 700 (ex 2606) which was superheated in 1913, at Leeds Holbeck, c1925. It was one of two of the class fitted with G & F Weir's feedwater heating system in 1914 and the pump can be seen on the left side of the smokebox. It was withdrawn two years later in November 1927. (W.H. Whitworth/MLS Collection)

No 700 seen at Holbeck from the other side showing the G & F Weir feedwater heater reservoir on the running plate, c1925. (J. Lord/MLS Collection)

No 706, which was built in 1901 as 801 and superheated in 1921, was fitted with a Dabeg feedwater heating system in 1930. It was withdrawn a few months after this photograph was taken, in May 1932. (MLS Collection)

No 759, superheated in 1923, seen at Nottingham, 29 March 1930. It has a six-wheel tender with coal rails. (W.L. Good/MLS Collection)

No 726 after a casual repair at Derby, June 1937. It was the last survivor of the class, withdrawn in September 1952. (W. Potter/ MLS Collection)

Two 'Belpaires', 773 and 739, stored at Hellifield on 11 October 1936. Both were repaired and returned to traffic, 773 being withdrawn in April 1940 and 739 not until December 1949. (MLS Collection)

No 729 stored at Sheffield Grimethorpe in June 1949. It was withdrawn a couple of years later in June 1951. (J. Davenport/MLS Collection)

No 40741 of Bristol Barrow Road renumbered and with 'BRITISH RAILWAYS' inscribed on the tender, c1949. It was withdrawn in October 1951. (Photomatic/MLS Collection)

No 40758 at Leeds after a casual repair, 30 April 1949. It was reboilered in 1947 and following a further light repair in December 1949, was withdrawn in March 1951 having achieved the highest recorded mileage of any of the 3Ps, 1,506,123. (H.C. Casserley/MLS Collection)

One of the last two survivors, 40728, at Derby, renumbered but the tender still showing 'LMS', c1950. It was withdrawn in July 1952. (MLS Collection)

it of low priority. However, the LMS chose 702 and 706 to be fitted with Dabeg feed pumps in 1929. A 10-12 per cent saving in coal and water had been anticipated but in trials showed only 5-7 per cent. No further engines were modified and these two engines were withdrawn in 1931 and 1932.

As stated earlier, the seven unrebuilt saturated boiler engines were withdrawn in 1925 and 1926. The first superheated engines to be withdrawn were three of the first ten, 700, 704 and 705 in 1927. Nos 703, 709, 732, 752, 760 and 766 followed in 1928 and 1929. Twenty-five were withdrawn in the 1930s and the following twenty-five were taken into BR ownership in 1948 – 711, 715, 720, 726, 727-728, 729, 731, 734-736, 738-741, 743, 745, 747, 748, 755-758, 762. However, only 40726, 40728, 40740, 40741, 40743, 40745, 40747 and 40758 lasted long enough to receive their British Railways number. No 40743 was the only one to receive the BR mixed traffic lining with BRITISH RAILWAYS on the tender, being withdrawn in July 1952. No 40726 was the last survivor, withdrawn just two months later.

Operations

The initial reason for the order was the need for more powerful engines with smaller diameter coupled wheels for the Peak District gradients between Derby and Manchester. However, the first five engines were allocated to Leeds from whence they were tested over all the Midland main lines – to Carlisle and London and Derby to Manchester. Nos 800 and 801 were allocated to Kentish Town, and 802-804 to Carlisle. The next batch were split between Leeds (2781-2783), Manchester (2784-2786) and Kentish Town (2787-2790). The allocation in 1902 was:

Manchester:	2606-2610
Carlisle:	800-804
Leicester:	2781, 2782
Kentish Town:	2783-2790

The prototype class 3, 2606, on a down express at Mill Hill, c1901.
(Locomotive & General/ MLS Collection)

No 2789 of the second batch of ten 'Belpaires' double-heading a class 2 4-4-0 on a down express at Mill Hill, c1903.
(Locomotive Publishing Co./ MLS Collection)

No 859 on test on the Derby-Bristol route at Cheltenham Lansdown station, c1904.
(MLS Collection)

After the delivery of all eighty engines in 1905 and the first allocation of the new Compounds, the depot allocation of the fleet was:

Manchester:	2606-2610, 823-834, 840-849, 860-869
Carlisle:	800-804, 810, 811
Leeds:	812-822
Derby:	835-839, 850-859
Leicester:	2781, 2782
Kentish Town:	2783-2790

With only the initial five Compounds in service by the beginning of 1905, the new 'Belpaires' were working all the principal expresses of the company, including the Derby-Bristol route, where the Compounds were not permitted until 1924.

The August 1904 *Railway Magazine* documented a number of runs with the Midland class 3s, before the 1907 renumbering. The writer, after experiencing the performance of a Johnson 'Single' on the new London-Manchester via Chinley express, returned from Leicester with what he described as the new 'Belpaire', 813 (later 723), with thirteen bogie coaches. After initial priming, the engine beat the 109 minute schedule for the 99 miles by half a minute (106 net) with a minimum of 41mph on the 1 in 132 four mile climb to Desborough and 43 on the 1 in 120 to Sharnbrook summit. It averaged nearly 50mph over the 18.5 miles from Bedford to Luton, mostly uphill at 1 in 200 to Leagrave summit. Then Kentish Town's 2783, built in 1902 and later renumbered 712, went north with twelve bogie vehicles and got to Leicester in 105 minutes 35 seconds (104 minutes net). Highlights were 42mph minimum at Sharnbrook, 45 at Desborough and 49 at Kibworth.

The same article reported the brand new 839 (749) taking over twelve bogie vehicles from a class 2 4-4-0 at Leicester and keeping time easily to Manchester, the hardest work being a sustained 33mph on the 1 in 100 before Millers Dale, followed by a rapid descent after Chinley through Buxworth. The recorder returned to Leicester with 827 (737) and ten bogie coaches in 113 minutes for the 97 heavily graded miles (108 net). Engines changed at Leicester to one of the five prototype Compounds, 2634 (later 1003).

The magazine took up the theme again the following month (the Midland and Great Western expresses were dominating the 'Locomotive Practice and Performance' articles at that time) and reported 830 (740) with 170 tons on the 4.10pm Manchester, running to Leicester in just 103 minutes 45 seconds including signal checks, with 36mph sustained on the 1 in 90 long climb to Peak Forest. Speed was not allowed to exceed 70mph on the descent through Millers Dale and Bakewell. No 830 then ran through to St Pancras without the usual change at Leicester, covering the 99 miles in nine seconds over 100 minutes, arriving five minutes before time, in 3 hours 20 minutes actual running time from Manchester.

The writer, Rous-Marten, then described a return non-stop trip to Sheffield and back (158 miles each way), with 821 (731) on the 9.45 St Pancras and 822 (732) on the 3.45pm from Leeds. Both trains loaded to 180 tons. The 158-mile non-stop down run to Sheffield was completed in 182 minutes 7 seconds, two minutes early. Kettering was passed in 79¼ minutes and Leicester in 110¾ – the need to conserve water on such a long run meant that the main climbs were taken easily, with speed dropping to 27mph at Sharnbrook and no more than 39 at Desborough. The return on the 3.45pm Leeds was more energetic as the train had left Carlisle nineteen minutes late reduced to nine minutes by the time the train left Leeds. Despite slowings to 15mph at Normanton, Tapton and Trent, St Pancras was reached in 3 hours 39¼ minutes from Leeds, the last leg from Leicester taking 108 minutes, the arrival in London being just four minutes late.

Rous-Marten devoted another article to the performance of the new Midland 4-4-0s in the September magazine. He stated near the beginning of the article that 'Mr Johnson's Belpaire engines are entirely capable in tackling the line's gradients and with their large double-bogie tenders, can run right through from London to Manchester without the traditional engine change at Leicester.' Then he quoted a run with Kentish Town's 2788 (717) on the 10am St Pancras with a train of 170 tons, giving the following outline details:

0	St Pancras	00.00	
6.9	Hendon	08.55	
	Elstree	14.28	
19.9	St Albans	20.59	
	Sandridge	-	signal stand of one minute costing a good three minutes' delay
30.3	Luton	33.29	
49.8	Bedford	48.50	77 (ave)
	Sharnbrook	-	55
61	MP 61	60.00	
72	Kettering	69.50	67½ (ave)
	Desborough	-	50/52
	Mkt Harboro'	-	50*
99.1	Leicester	99.09	(95½ net)
		00.00	
	Trent	19.09	
	Derby	29.00	
	Peak Forest	-	36 (after 6 miles 1 in 90 rising)
			Signal checks approaching Manchester

Manchester was reached in 3 hours 31 minutes and 48 seconds from London, 3 hours 20 minutes actual running time and 3¼ hours net. He fails to mention the maximum speeds reached, being more interested in the performance on the banks, although it is clear from the above times that 2788 averaged 77mph on the Luton-Bedford stretch and there speeds well in excess of 80mph were probable. A few months later, when he still seems to hanker for the days of the Johnson 'Singles', he quotes a non-stop St Pancras-Nottingham run with 2788 again, which clearly ran very fast, passing Bedford in 46 minutes net and averaging nearly 75mph between St Albans and Bedford, which again indicates to me that after Leagrave, speeds in the high eighties were probable. The return from Nottingham was equally impressive. Class 2 4-4-0 2428 lost six minutes between Leeds and Nottingham and 839 (749) took over, departing eight minutes late with 220 tons. It passed Kettering (51¾ miles) in 52 minutes 54 seconds, surmounted Sharnbrook at 47½mph and averaged 78mph on the descent to Bedford. It sustained almost 60mph on the long 1 in 200 to Leagrave and averaged 75mph from Radlett to Hendon after a special unscheduled halt at Luton. Net time for the 123 miles non-stop was 124 minutes.

In September 1905, however, Rous-Marten experienced 2782 (711) with 280 tons on a St Pancras-Leicester 105 minute schedule, and found the load was too great to keep time, the net time being 108 minutes – the maximum permitted load for the schedule was only 170 tons. No 847 (757) had an even heavier load – 300 tons – but dropped ten minutes on the 105 minute schedule (109 net). Fortunately, most trains were restricted to the scheduled load or received a pilot locomotive, usually a Johnson 4-2-2.

In 1908, after the renumbering, the allocation had changed to:

Kentish Town:	709-711 (804, 2781, 2782), 713-719 (2784-2790)
Leicester:	700-704 (2606-2610)
Carlisle:	720 (810)
Leeds:	712 (2783), 721-742 (811-832)
Manchester:	705-708 (800-803), 743-750 (833-840), 776-779 (866-869)
Derby:	751-773 (841-863)
Nottingham:	774, 775 (864, 865)

R.E. Charlewood, in a *Railway Magazine* article of 1909, quotes a couple of logs of class 3 engine performances on the Settle & Carlisle section:

Hellifield-Carlisle, c1907
730 (ex 840)
240 tons

Miles	Location	Times	Speeds	Gradients
0	Hellifield	00.00		
0.8	MP 232	01.58		1/181 F
3.4	Settle Junction	-	sigs	
5.3	Settle	08.12		1/100 R
11.3	Horton-in-Ribblesdale	19.05	30½	1/100 R
17.3	Blea Moor	30.06	31½	1/100 R
22.3	Dent	36.44		
28.5	Ais Gill	43.31	45	1/165 R
35.5	Kirkby Stephen	49.38		1/100 F
44.3	MP 275 (Ormside)	56.17	80½	1/100 F
46.7	Appleby	58.06		
54.1	Culgaith	64.36	68	
61.7	Lazonby	71.43	64	
67.3	Armathwaite	77.09		
74.5	Scotby	82.58	74/sigs	1/132 F
76.8	Carlisle	86.50		

Carlisle-Hellifield
10.30am Edinburgh-St Pancras
705 (ex 800)
255 tons

Miles	Location	Times	Speeds	Gradients
0	Carlisle	00.00		
2.7	Scotby	05.55		1/132 R
8	MP 300 (Cotehill)	14.16	38	1/132 R
9.8	Armathwaite	16.27		1/132 F
15.4	Lazonby	22.22		
23.3	Culgaith	30.27	58	
30.8	Appleby	39.10		
33	MP 275 (Ormside)	41.36	36½	1/100 R
41.8	Kirkby Stephen	54.05	32½	1/100 R

		Carlisle-Hellifield		
		10.30am Edinburgh-St Pancras		
		705 (ex 800)		
		255 tons		
Miles	Location	Times	Speeds	Gradients
48.3	Ais Gill	65.58	33½	1/100 R
54.6	Dent	73.08		
59.5	Blea Moor	78.28		
65.5	Horton-in-Ribblesdale	83.51	72	1/100 F
71.5	Settle	88.57	72	1/100 F
76.8	Hellifield	94.39	½ E	

An article in the May 1911 *Railway Magazine* tables the log of a class 3 'Belpaire' 4-4-0 on the 162 mile non-stop run from Rotherham to St Pancras with the 1pm Leeds express allowed 177 minutes, including pass to stop of 104 minutes in from Leicester.

		Rotherham-St Pancras		
		735		
		205 tons		
Miles	Location	Times	Speeds	Gradients
0	Rotherham	00.00	7 L	
7.8	Killamarsh	10.05	62½	
16	Chesterfield	18.20		
23.5	Does Hill	28.55	42	1/160 R
42	Trent Junction	49.40	sigs	colliery slacks
58.1	Syston	-	65½	
62.8	Leicester	72.30	4½ L	
72.8	Kibworth	-	50	1/199 R
78.8	Market Harborough	-	71½	1/167 F
82.3	Desborough	-	42	1/132 R
89.8	Kettering	103.25	76½	1/118 F
105	Sharnbrook	-	47½	1/120 R
112	Bedford	123.45	80½	1/119 F
129.1	Leagrave	-	51½	1/200 R
131.5	Luton	144.40	60	
143	St Albans	155.05	easy	
146.9	Radlett	-	77½	1/176 F
149.6	Elstree	-	55½	1/200 R
155.1	Hendon	-	71½	1/176 F
162	St Pancras	175.45	(173 net) 3¾ L	

On another run on the same train, 745 with 180 tons took exactly the same time to Leicester and, although it had a lighter load, it was encountering a very strong side wind. It then fell only from 60 to 54½mph at Kibworth, touched 75 at Market Harborough, fell to 47 at Desborough summit, flew through Kettering at 77½, climbed to Sharnbrook with a minimum speed of 45½mph and speed had risen to 77 ½mph on the descent when disaster struck and the engine suffered the breakage of a big-end brass, coming to a stand at Bedford having traversed the 112 difficult miles in 123 minutes. A Kirtley 2-4-0 replaced the 4-4-0 and ran to London in 57 minutes 10 seconds dropping only two minutes on the schedule, excluding the time taken for the engine change.

In 1909 and 1910, there were comparative tests between the class 3 'Belpaires', the class 4 '999s' and the 'Compounds'. Seventy-six runs were involved in the trial on Leeds-London and London-Leeds expresses, with loads of around 180 tons plus passengers. Nos 713 of Kentish Town and 733 of Leeds were the class 3s selected, the main purpose being to compare coal consumption figures, the '999 ' class did the work on 96 per cent of the class 3's coal consumption figures and the 'Compound' on 93 per cent. There were tests with 'H' boilered class 2 4-4-0s later in 1910, and these showed a coal consumption greater than that of the 'Belpaires' by 10 per cent.

C.J. Allen in *Railway Magazine* articles around 1913/14 quoted a number of runs with both 'Belpaires' and 'Compounds', a selection of the class 3 runs are tabled below.

Derby-Manchester, c1910

10.25am St Pancras-Manchester

Miles	Location	726 187/200 tons Times	Speeds		725 187/200 tons Times	Speeds		Gradients
0	Derby	00.00			00.00			
5.3	Duffield	08.45			08.15	60		
10.4	Ambergate	14.35	sigs	¼ E	14.00	28*	1 E	
17.2	Matlock	23.05			22.15			1/177 R
21.6	Rowsley	27.50	56½	1¼ E	26.35	64	2½ E	
25	Bakewell	32.25			30.45			1/102 R
27.3	Longstone	36.10			34.20			1/100 R
31.4	Millers Dale	42.20	35½	1¾ E	40.30	41/pws	3½ E	1/100 R
36	Peak Forest	50.25	29½	2½ E	49.25	31½	3½ E	1/90 R
41.6	Chinley	56.40		3¼ E	56.10	55*	3¾ E	1/90 F
44.3	New Mills S. Jcn	59.20		2¾ E	58.35		3½ E	1/89 F
53.4	Cheadle Heath	68.00		3 E	67.00	74	4 E	1/100 F
59.2	Chorlton	-	sigs/sigs		77.50	sigs/sigs	¼ E	
61.4	Manchester	82.00	(79 net)		85.00	976½ net)	T	

		Manchester-Derby, c1910					
		706			727		
		205/215 tons			212/225 tons		
Miles	Location	Times	Speeds		Times	Speeds	Gradients
0	Manchester	00.00			00.00		
4.2	Chorlton	05.45	sigs	1¼ E	07.40	sigs stand	¾ L
8	Cheadle Heath	13.25		½ L	13.40		1¾ L
		00.00					
12	Hazel Grove	08.35			20.00	33	1/100 R
17.1	New Mills S. Jcn	16.30	38	2½ L	28.35	35	2½ L	1/100 R
19.8	Chinley	21.25	34		33.45	30 ½	1/90 R
21.7	Chapel-en-le-Frith	25.10	37		37.35	30	1/90 R
25.4	Peak Forest	32.15	30	3¼ L	45.00	28	2 L	1/90 R
30	Monsal Dale	36.50	70	2¾ L	50.00	65	2 L	1/90 F / 1/100 F
36.4	Bakewell	42.45	67		56.30	60	1/100 F
39.8	Rowsley	45.55	pws	2L	59.45	easy	1¾ L	1/100F / 1/184 F
44.2	Matlock	51.00			64.50		
51	Ambergate	58.55		3 L	72.05		1 L
56.1	Duffield	64.25	68		78.25		
60.6	Nottingham Road	68.35	60		82.35	60	
61.4	Derby	70.00	(67 net)	2 L	84.00	(82 net)	1 L

		Leicester-St Pancras, c1912					
		737 - Leeds			744 - Manchester		
		195 tons			195 tons (*170 from Kettering)		
Miles	Location	Times	Speeds		Times	Speeds	Gradients
0	Leicester	00.00			00.00 Diverted from Nottingham (floods)		
3.7	Wigston	05.20	55½		05.55		
7.3	Great Glen	-	53		-		1/199 R
10.2	Kibworth	12.45	74		13.30		
16	Market Harborough	18.05	*		19.00		1/238 F
20.8	Desborough	23.55	45		25.05		1/132 R
27	Kettering	29.10	79	¼ L	30.45 (coach slipped)	1¾ L	1/118 F
34	Wellingborough	34.55		1 E	37.05	1 L	

		Leicester-St Pancras, c1912					
		737 - Leeds			744 - Manchester		
		195 tons			195 tons (*170 from Kettering)		
Miles	Location	Times	Speeds		Times	Speeds	Gradients
36.3	Irchester	-	pws		-		
39.3	Sharnbrook summit	42.00	40		42.45		1/120 R
42.3	Sharnbrook	-	79		-		1/119 F
49.2	Bedford	51.10	68	¼ L	51.25	½ L	
57.3	Ampthill	60.10			59.40		1/200 R
66.3	Leagrave	70.55			69.50		1/200 R
68.8	Luton	73.20	70	1¼ L	72.20	¼ L	
79.2	St Albans	82.50		¾ L	82.25	½ L	
83.8	Radlett	-	77½		-		1/176 F
86.5	Elstree	89.30			88.50		1/200 R
92	Hendon	94.50	70		93.35		
	Finchley Road	98.10			96.55		
99	St Pancras	103.45		¼ E	102.45	1¼ E	

As the Derby-Bristol route could not then be allocated 'Compounds' for the most prestigious trains, it is worth 'quoting' a run on the Bristol Mail with a class 3, on an unknown date:

		Cheltenham-Birmingham New Street		
		761 -Derby		
		285/305 tons		
		7.20pm Bristol Mail		
Miles	Location	Times	Speeds	Gradients
0	Cheltenham	00.00		
3.8	Cleeve	06.05		
7.2	Ashchurch	09.35	64	1/295 F
9.4	Bredon	11.50	53½	1/301 R
13.2	Defford	15.55	59	L
18	Abbots Wood Jcn	21.30	47½	1/301 R
20.4	Spetchley	24.30		
24.6	Dunhampstead	29.20	56	L
31.3	Bromsgrove	37.50		
0		00.00 Attach 'Big Bertha' 0-10-0		
2.2	Blackwell	07.30	17½ (ave)	1/37¾ R

Cheltenham-Birmingham New Street
761 -Derby
285/305 tons
7.20pm Bristol Mail

Miles	Location	Times	Speeds	Gradients
3.6	Barnt Green	10.05		
8.7	King's Norton	16.25		1/301 F
10.9	Selly Oak	19.45		
<u>14.2</u>	<u>Birmingham</u>	<u>25.55</u>		

By 1914, the Nottingham allocation had been augmented by seven more, 720 from Carlisle and 720-723, 730, 733 and 738 from Leeds, as the Settle & Carlisle line was in the hands of the class 4 '999' class and the 'Compounds' by then. The allocation remained the same through the First World War, but by 1920 Leicester had acquired 707, 710, 711, 713, 750 and 764 from Manchester, Kentish Town and Derby, but Derby then replaced the Kentish Town loss with 752, 753, 758 and 762. Derby's 771 and 772 went to Leeds leaving Derby with fifteen, 751, 754-757, 759-761, 763, 765-770, a loss of eight engines.

No 767 with a Deeley 'face', paired with a six-wheel tender obtained from sister engine 761, on a down express at Elstree, c1910. (F. Moore/MLS Collection)

Leicester's 702 (the former 2608) with a down express of bogie rolling stock at Elstree, c1912. (MLS Collection)

Derby's 753 on a down express composed of six-wheel and bogie coaches, leaving Elstree Tunnel, c1912. (MLS Collection)

The Midland Class 3 • 147

Manchester's 749 finds itself on freight duties at Elstree at the beginning of the First World War, c1914. (H. Gordon Tidey/MLS Collection)

No 763, converted to oil-burning during the 1921 miners' strike, alongside 'H' boilered 4-4-0 371 at Derby, 9 August 1921. (H.C. Casserley/MLS Collection)

A scene at St Pancras station with two class 3s, the saturated boiler 738 waiting to depart on a down express and the superheated 715, 30 July 1921. (MLS Collection)

The superheated 777 near Chapel-en-le-Frith, c1922. (W.L. Good/ MLS Collection)

No 710 in Aintree Sidings after arriving with a Grand National excursion train, 1922. (E.M. Johnson Collection)

Oil-burning 773 with an up express near Ratcliffe Junction, 1921.
(MLS Collection)

At the Grouping, the allocation remained stable until 1925 when 765 was tested from Camden shed. Tests were also undertaken with 701 on the former LT&SR with a '483' class 562. The class 2 had the slightly lower coal consumption figure but that led to no significant change there. The 'Belpaires' had been and remained the mainstay of the Derby-Bristol route as heavier engines like the Compounds were not allowed there until after 1924. The building of the LMS 'Compounds' in significant numbers led to a major redistribution of the remaining 3Ps (withdrawals of the saturated boiler engines had begun). The 3Ps in turn displaced many of the 2P 4-4-0s. The new depot allocations around 1927 were:

Gloucester:	715-717, 750-752
Bristol:	718
Bedford:	719, 753-755, 763, 764
Sheffield:	727-731
Saltley:	712-714

Further moves in 1930 included in addition to the above:

Bedford:	720, 750 (from Gloucester), 760, 762, 765, 767, 768
Lancaster:	734-736 (+747 in 1933)
Skipton:	773
York:	727-729 (from Sheffield)

The Midland Class 3 • 151

No 776, the engine fitted experimentally with a Kylala blastpipe, near Milford Tunnel on the Derby-Manchester line through the High Peaks, c1925. (MLS Collection)

708 storming out of Poulton-le-Fylde with what looks like an excursion from Blackpool, April 1925. (A.G. Ellis/MLS Collection)

No 746 on a Blackpool-Manchester train near Farington Grove Junction, 1925. (MLS Collection)

Bedford's 755 waits to depart from St Pancras station with the 5.05pm Sunday semi-fast train for Bedford, 1930. (G.A. Coltas/MLS Collection)

No 759 piloting a Midland 2P near Ambergate on a Manchester express, c1930. (Ellison/MLS Collection)

No 745 stopping at Chinley station with the 8.45am Manchester Central-Sheffield train, 13 September 1931. (A.W. Croughton/MLS Collection)

No 748 near Chinley with a Manchester-Sheffield stopping train, 2 April 1934.
(MLS Collection)

Leicester's 711 piloting Stanier 'Black 5' 5276 on a Manchester-St Pancras express, 7 August 1937.
(R.D. Pollard/MLS Collection)

One of the Sheffield engines was timed on a Manchester-Liverpool train in 1931, just before being transferred to the LMS sub-shed at York. By this time there were few logs of the LMS 3Ps as their work was now mainly secondary in nature covering Sheffield – Manchester and St Pancras-Bedford semi-fast and stopping services.

Manchester Victoria-Liverpool Exchange, 10 January 1931

729 - Sheffield

8 chs, 230 tons

Miles	Location	Times	Speeds	Gradients
0	Manchester Vict.	00.00	18 L	
0.7	Salford Central	03.00		
2.3	Pendleton	05.25	53	
4.4	Pendlebury	09.15	32	1/99, 1/80 R
7.5	Walkden	13.55	50	
11	Atherton	17.55	58	1/150 F
14.7	Crows Nest Jcn	22.30	pws 35*	
15.2	Hindley	23.15	pws 30*	
18	Westwood Park	26.35	60	1/150 F
19.3	Pemberton	28.45	31	1/92 R
21	Orrell	31.25	38	1/92 R, 1/282 R
29.5	Kirkby	40.35	70	1/118 F, 1/316 F
31.4	Fazakerley	42.20	65	L
35	Sandhills	46.05	55	
36.5	Liverpool Exch	49.15	22¼ L	

During the Second World War, the remaining twenty-five locomotives were retained for freight and banking work, six going to Rowsley on the Derby-Manchester line (711, 738-741 and 773, the latter replaced by 757 in 1941). Nos 726 and 727 were shedded at Canklow, more were allocated to Sheffield depot for goods train working. A number of them appeared in London during the war on goods train duties from the East Midlands.

After the war, 756 was at Peterborough, and 741 was at Bristol, being lent for test purposes to Templecombe on the Somerset & Dorset in 1950. Some worked from Kettering on seasonal traffic in 1946/7. Bedford retained a number for semi-fast trains to St Pancras and cross-country stopping trains to Oxford. Many were withdrawn in the immediate post-war period before nationalisation, the depot allocation of the last survivors being:

Nottingham's **774** on a summer Saturday relief express at Peak Forest, August 1938. (E.R. Marten/MLS Collection)

40726 – Canklow (withdrawn September 1952)

40728 – Sheffield (withdrawn July 1952)

40729 – Sheffield (withdrawn June 1951)

40741 – Bristol (withdrawn September 1951)

40743 – Leeds, transferred to Bedford in March 1951 (withdrawn July 1952)

40747 – Leeds (withdrawn June 1951)

40758 – Leeds (transferred to Bedford February 1951 and withdrawn in March)

40762 – Bedford (withdrawn February 1951)

A final 'Stephenson Locomotive Society' special train was worked by 40726 from Manchester to Hull and back on 24 August 1952 and returned from Manchester to Sheffield on the 4.03pm local train the following day, after which it was withdrawn.

No 759 on a southbound troop train on the West Coast main line near Barton & Broughton, 24 August 1940. No 759 was withdrawn at the end of the war in February 1946. (MLS Collection)

No 40728 pilots one of Bristol's 'Jubilees' 45682 *Trafalgar* on the 7.40am Bristol-Newcastle express at Derby station, 11 August 1950. (H.J. Buckley/MLS Collection)

The Midland Class 3 • 157

The last survivor, 40726, waits to depart from Manchester Central with the Stephenson Locomotive Society special train for Hull, 24 August 1952.
(MLS Collection)

No 40726 on the SLS special train at Chinley en route to Hull, 24 August 1952.
(A.C. Gilbert/MLS Collection)

Chapter 4
THE MIDLAND CLASS 4 (COMPOUND & SIMPLE)

The Class 4 Compound
Design & construction

The design of compound locomotives in the UK had had a very chequered career. The LNWR Webb compounds had been very undistinguished and Worsdell's two-cylinder compound 4-4-0s for the North Eastern Railway were equally unsuccessful. Worsdell's assistant, W.M. Smith, then persuaded Worsdell to rebuild one of the 4-4-0s, No 1619, as a three cylinder compound with a single high pressure cylinder inside the frame with the exhaust from that cylinder flowing to two low pressure outside cylinders. This was markedly better than in its previous guise and although no more NE engines were converted in this way, Smith was able to share the experience with Johnson whom he had known for many years. With the increasing weight of trains on the Midland Railway and the constantly developing services, Johnson used authority previously given to construct two compound 4-4-0s using the Smith principles. Unique to these two engines was the ability to adjust the settings of the high-pressure and low-pressure valve gears independently.

The two engines, numbered 2631 and 2632, were constructed during the autumn of 1901, tested and entered traffic in January 1902. Their key dimensions were:

Cylinder diameter	
– high pressure	19in x 26in
– low pressure	21in x 26in
Stephenson's Link	Motion
Coupled wheel diameter	7ft 0in
Bogie wheel diameter	3ft 6½in
Boiler pressure	195lbs psi
Heating surface	1,598sq ft
Grate area	26sq ft
Axleload	19½ tons
Weight	
– Engine	59 tons 10 cwt
– Tender	52 tons 12 cwt
– Total	112 tons 2 cwt
Water capacity	4,500 gallons
Coal capacity	5 tons
Tractive effort	19,110lbs

The weight, and particularly the 19½ ton axleload, caused major concern and debate at the Midland Railway Board, the company's civil engineer complaining that the engine exceeded the axleweight agreed in 1897 for the increased power engines (the class 3 'Belpaires') and that 108 bridges would have to be strengthened at a cost of £96,000 if the weight of the new engines could not be altered by rebalancing. As a result of this, the two new Compounds were restricted to the Leeds-Carlisle route until the bridge strengthening had been carried out. Johnson had pleaded that the stresses produced by the hammer blow of a three-cylinder engine on the track mitigated the impact of the heavier engine and the civil engineer responded with permission to go to 19 tons 1 cwt, but not the 19½ tons of 2631 and 2632.

The trials of the two Compounds over the Settle and Carlisle route proved highly satisfactory and the remaining three engines of the order, 2633-2635, were constructed between July and November 1903. The only significant change from the two prototypes was that the independent control of the high- and low-pressure admission of steam to the cylinders was abandoned as the need to work

The Midland Class 4 (Compound & Simple) • 159

The second Smith/Johnson Compound, 2632, as built in 1902, in Works grey for official photograph. Note the large 4,500 gallon bogie tender. (F. Moore/MLS Collection)

The reverse (left hand side) of the second Smith/Johnson Compound, 2632, as built in 1902, in Works grey. (MLS Collection)

The first Smith/Johnson Compound, 2631, undergoing initial tests on the Leeds-Carlisle line, 1902. (W.L. Good/MLS Collection)

Smith/Johnson Compound 2634, as built in 1903. Note the straight running plate compared with the first two's raised plate over the cylinders. (Locomotive Publishing Co./MLS Collection)

them independently had not been necessary in the trials. By this time, bridges had been strengthened and the Compounds were then accepted over the company's main routes apart from the Derby-Bristol line, where the class 3 'Belpaires' were the maximum axleload engines permitted until LMS days.

Johnson retired at the end of 1903 and his successor, Deeley, was sufficiently impressed with the five Compounds' performance to negotiate a deal with Smith to cover his patent, and he got the authority to build forty more to a modified design. The first batch, numbered 1000-1009, were constructed in the autumn of 1905, twenty more, 1010-1029, during 1906 and the final ten in 1908/9. The first five were renumbered 1000-1004 in 1907 and 1000-1029 became 1005-1034. The 1908 batch were numbered 1035-1044 from the start. There were some significant dimension changes, the boiler pressure being raised to 220lbs psi, grate area to 28.4sq ft, but total heating surface reduced to 1,485sq ft. Deeley also simplified the controls and the external appearance of the Deeley engines was different with a more substantial cab, a more harmonious treatment of the rear coupled wheel splasher integrated into the cab design and a six-wheel rather than bogie tender following the introduction of water troughs on the main routes. Another external difference was the shape of the smokebox from 1010 (1015 post 1907) which was circular rather than extended at the bottom to the saddle as on the Johnson engines. Their

No 1013 of the first batch of Deeley Compounds after renumbering from 1008 at Kentish Town, c1907. Note that the Deeley engines had a raised running plate over the coupled wheels.
(E.M. Johnson Collection)

No 1014 of the second batch of Deeley Compounds, later renumbered 1019, with circular shaped smokebox as fitted from 1010 onwards, at Leeds Wellington, 1906. (MLS Collection)

No 1003, formerly Smith/Johnson Compound 2634, built in 1903, renumbered in 1907, fitted with Deeley smokebox, chimney and rebuilt bogie to six-wheel tender. It is seen at Nottingham, c1912. It was superheated in 1915. Note the straight running plate of the former 2633-2635 apparent still here. (F.H. Gillford/MLS Collection)

livery was simplified from the fully lined and decorated Johnson red of 2631-2635 to a revised red livery with crest on the cabside, smokebox numberplate, large numerals and reduced lining on the tender.

Failures in the complex Smith valve starting arrangement on the five Johnson engines was overcome by replacing with the Deeley regulator and gradually other changes were made to bring them in line with the Deeley Compounds. In 1911, they were authorised for rebuilding with new cylinders and boilers and therefore would become identical to the later engines. Their bogie tenders had been replaced by six-wheel tenders after 1908, some by rebuilding of the bogie tender, others by exchange with engines of other classes.

The next major change was the decision to fit one of the Deeley engines with a superheater, which had been proven to be successful with the earlier conversion of the 'Belpaires' and many of the Johnson class 2 4-4-0s. No 1040 was fitted with a superheater in 1913 and its G9 boiler was redesignated G9S. The success of this led to a change in the authority for 1000-1004 to include new frames and a new boiler with Schmidt superheater. All except 1002 were rebuilt in 1914 and 1915. No 1002 was not returned to traffic after rebuilding until 1919. During the war, no further Compounds were superheated, priority being given to the class 2 and 3 4-4-0s. Nos 1005-1039 and 1041-1044 only received superheated boilers when their existing G9s required replacement,

No 1017 with round smokebox at Leicester station, with proud driver posing on the running board, 1912. (MLS Collection)

No 1000 as rebuilt in 1914 with new frames and superheated boiler, in Works grey. (Railway Photographs/MLS Collection)

The first Deeley Compound to be superheated, 1040, in 1913. It is posing here at Kentish Town shortly afterwards, 1914. (MLS Collection)

1008 and 1014 being the first in 1919 and 1022 the last in 1928.

Variations in boiler pressure took place during the First World War with some downgrading to 190lbs psi, then finally all were brought in line to 200lbs psi, a pressure that was standard for the Compound superheated boiler. The Compounds were complex machines and although their performance on the road was superior to the other Midland passenger engines, their maintenance costs were high. Their average mileage between shop repairs, assessed in a report of 1910, was only 34,238 miles, compared with 52,294 for the 'Belpaires', and their time in Derby Works at overhaul was on average three weeks longer. Although their weekly mileage was higher (970 to 810 miles), their average availability was 64.8 per cent compared to the Belpaires' 78.5 per cent, the latter being a comparatively good figure at the time when maintenance standards were high.

The changed dimensions for the final form of the Midland Compounds as taken over by the LMS in 1923 were:

Boiler pressure	200lbs psi
Grate area	28.4sq ft
Heating surface	1,681sq ft (incl superheater 360sq ft)
Axleload	19¾ tons
Weight	
– Engine	61 tons 14 cwt
– Tender	45 tons 18 cwt
– Total	107 tons 12 cwt
Water capacity	3,500 gallons
Coal capacity	7 tons
Tractive effort	21,840lbs

The Compounds were not included in the oil-burning conversion programme of 1921, but seven Midland engines were fitted with 450 gallon tanks in 1926 (1000, 1005, 1008, 1011, 1036, 1038 and 1041). A further three (1025, 1032 and 1043) received larger 500 gallon fuel tanks. All oil-burning equipment had been removed by the end of the year.

The Compounds were the only Midland engines to retain the red livery after the Grouping and did so until wartime black was applied from 1940. BR lined mixed traffic livery was used after nationalisation for a few (41005, 41006, 41021, 41035, 41038 and 41044) and 41025 had the odd combination of lining but LMS retained on the tender. Most remained plain black with LMS or BRITISH RAILWAYS on the tender.

The Compounds were displaced from most of their best turns by the Stanier 'Black 5s' in the 1930s and efficiency fell away, but no withdrawals took place before 1945. In fact, all were taken into BR ownership, though 1029 was withdrawn in January 1948. As late as 1947, seven engines had received General repairs and three had involved fitting new boilers. Eight more received Works repairs

No 1005 converted to oil burning with two 450 gallon fuel tanks in 1926 for the duration of the General Strike, at Derby, 1926. It has the Fowler short chimney.
(MLS Collection)

No 1000 after overhaul at Derby and repainting in LMS black livery before re-entering traffic, c1928. It was rebuilt with superheated boiler in 1914. (MLS Collection)

No 1007 at its home depot, Bedford, c1935. (MLS Collection)

The Midland Class 4 (Compound & Simple) • 167

No 1009, also at Bedford, but seen from the left hand side. Note the mechanical lubricator on the running plate.
(L. Thomsett/MLS Collection)

No 41016 in shabby condition shortly after nationalisation, locomotive renumbered but tender still bears LMS lettering. It is seen passing Chinley, 4 September 1949.
(J.D. Darby/MLS Collection)

No 41009, one of the few Midland Compounds to be painted fully in the BR mixed lined traffic livery, at Bedford, 19 March 1949.
(H.C. Casserley/MLS Collection)

in 1948, but more were withdrawn. Four received General repairs in 1949 and four more (the last) in 1950. Three (41025, 41041 and 41044) had Intermediate repairs in 1951. Most of the former Midland Compounds were withdrawn in 1951 and 1952, the last survivor being 41025 withdrawn in January 1953. However, 41000 was set aside at Derby in September 1951 pending a preservation decision.

Operations
The first two Smith/Johnson Compounds, 2631 and 2632, were allocated initially to the Settle & Carlisle line as their weight prohibited them from operating over the other Midland main line until bridge strengthening had been carried out. No 2631 was based at Leeds and 2632 at Carlisle. Both engines were tested rigorously (see the photograph of 2631 on pages 160 and 180). No 2631 with 229 tons on the 5.33pm from Leeds on 5 September 1902 completed the 76.7 miles from Hellifield to Carlisle in 79 minutes with 39/44 at Helwith Bridge, 39 at Horton-in-Ribblesdale, 35mph minimum at Ribblehead and 37 at Blea Moor. It then touched 86mph on the steep descent after Kirkby Stephen at Ormside. The same engine on the 4pm from Carlisle on 12 October with 248 tons started the climb from Appleby at 66mph, fell to 42 on the first 1 in 100 to Crosby Garrett, recovered to 50 on the slight easing before Smardale Viaduct, sustained 40/37 on the long 1 in 100 past Kirkby Stephen to Birkett Tunnel and cleared Mallerstang at 43 and the summit at Ais Gill at 37mph. A maximum indicated horsepower of 1,000 was calculated when the engine was working on full regulator on the last stretch to Ais Gill with cut-off 63 per cent on the high pressure and 57 per cent on the low pressure cylinders (these cut-offs are equivalent to about 25 per cent on a simple engine). Coal consumption was considered very satisfactory ranging from 25lbs per mile when hauling

150 tons to 45lbs per mile on a faster run with 250 tons. The engines were free running too. The 86mph at Ormside on the run quoted above was not exceptional and speeds of around 85mph were achieved on the descent from Ribblehead on the up runs.

Charles Rous-Marten timed sister engine 2632 on the 11.30am St Pancras with 250 tons which sustained 36-40mph on the climb to Ribblehead and reached 88mph at Ormside. He timed 2632 again on an up run when it beat 2631's best test run, with 240 tons, reaching 45mph on the first section of 1 in 100, 53 at Smardale Viaduct and 43 at Ais Gill summit. It then reached the astonishing speed for that time and that route of 92mph for over two miles on the descent to Settle. This is, as far as is known, the highest speed ever achieved by a Midland Compound. In regular service later, the Compounds were able to maintain time with comparatively light loads by fast hill climbing and moderate downhill speeds.

Of the next three engines built in 1903, 2633 went to Nottingham and 2634 and 2635 to Kentish Town as the route prohibitions had by then been removed. Rous-Marten again took special interest in these engines and recorded 2634 taking 350 tons from St Pancras to Leicester, 99 miles in 110¼ minutes including a stop at Bedford. From the standing start at Bedford to clear Sharnbrook summit at 40mph with this load was excellent as were 43 at Desborough and 48½ at Kibworth. After their exposure and testing on all the main Midland routes (apart from Derby-Bristol), these first five compounds were based at Leeds to work the Settle & Carlisle line.

The first ten Deeley compounds built in 1905, 1000-1009, were allocated to Kentish Town and the next 1906 built batch went to Nottingham. At the completion of the construction of the Deeley engines, the 45 compounds were allocated as follows (using post 1907 numbers):

Kentish Town:	1005-1015, 1042-1044
Nottingham:	1016-1026
Leeds:	1000-1004, 1027-1036
Manchester:	1037-1041

The compounds soon assumed the majority of Midland express working between London, Manchester, Nottingham, Derby, Leeds and Carlisle. A typical run from an early *Railway Magazine* article recorded the run tabled on the next page. It was undated but was probably around 1908 after renumbering with the engine one of the first ten of the Deeley engines, initially 1003.

No 2633 (later 1002) 'built in March 1903' on a down Manchester express near Ambergate, c1903. (F. Moore/MLS Collection)

St Pancras – Leicester
1008
229/245 tons

Miles	Location	Times	Speeds		Gradients
0	St Pancras	00.00			
1.5	Kentish Town	03.50		¼ E	1/178 R
6.9	Hendon	11.30		1½ L	1/200 F
12.4	Elstree	18.00	46		1/176 R
15.2	Radlett	20.55	67		1/200 F
19.9	St Albans	25.55	46½	1 L	1/176 R
24.6	Harpenden	31.30			
30.2	Luton	37.05	66	1 L	
32.8	Leagrave	39.30			
37.3	Harlington	43.35	79		1/200 F
41.8	Ampthill	47.10			1/200 F
49.8	Bedford	53.45	79	¼ E	
56.7	Sharnbrook	60.15			1/119 R
59.7	MP 59 ¾	64.25	39½	½ E	1/119 R
65	Wellingborough	69.25	70*	½ E	1/120 F
72	Kettering	76.35		½ E	
78.5	Desborough North	84.45	44½		1/136 R
82.9	Mkt Harborough	89.00	70*	1 E	1/132 F
86.3	East Langton	92.25	64½		
89.7	Kibworth North	96.10	48		1/130 R
95.4	Wigston	101.40	66*		1/199 F
99.1	Leicester	106.35		1½ E	

A number of further tests took place around this time. The North British initiated some tests with their Reid Atlantics, comparing them with an LNWR 'Experiment' 4-6-0 between Preston and Carlisle and with compound 1032 and an NER Atlantic between Carlisle and Edinburgh via the Waverley route. The Midland engine, 1032, kept time comfortably with 240 tons, though I cannot find how this compared with the other engines tested. In 1910, the Midland held comparative trials between the class 3 'Belpaires', the Deeley compound and the new simple class 4 '999' class to be described in the next section of the book (pages 194 to 203). Nos 1010 and 1034 were the two compounds tested on London-Leeds expresses with loads varying from 175 to 300 tons. The compounds, still in saturated form, showed the best economy, nearly 10 per cent better than the 'Belpaires' and around 5 per cent better than the simple class 4s. Further tests after the superheating of 1040 in 1913 showed significant advantage over both the saturated compounds and superheated 999s (the latter being better than the saturated compounds). A paper to the Institute of Civil Engineers by Henry Fowler in 1914 showed very clearly the benefits of compounding and superheating, some 35 per cent saving or more for both coal and water consumption of which two thirds was due to superheating. It is all the more remarkable, therefore, that apart from 1040 and the rebuilding of the Johnson compounds, nothing was done to superheat the other Deeley compounds until well after the First World War, most being superheated in LMS days. Priority was given to the superheating of the numerous class 2 4-4-0s. One can only assume that their improvement was more urgent in that the compounds, although they could be much better, were already more than satisfactory.

There were a number of 'British Locomotive Practice and Performance' articles in the *Railway Magazines* of 1913 and 1914 which gave details of good running by the Midland compounds, comments being made on the ease with which these engines met all the demands of the timetable. Although the scheduled times were comparable with the best Great Western and East and West Coast average speeds, since Paget's timetable and train control revision in 1909, loads on Midland expresses had been severely restricted so that there was no need for the hard running that for instance characterised the 'thrashing' of LNWR 4-4-0s to maintain time with 400+tons. If loads exceeded the maximum laid down pilot engines were provided. Some examples of good typical performances by the saturated Deeley compounds are tabled next.

St Pancras-Leicester, c1910

Miles	Location	1021 220 tons Times	Speeds		1039 155 tons Times	Speeds		1036 235 tons Times	Speeds		Gradients
0	St Pancras	00.00			00.00			00.00			
1.5	Kentish Town	04.15			03.25			05.35			1/178 R
6.9	Hendon	11.25			09.45			11.55			1/200 F
12.4	Elstree	17.25	50		15.05	60		18.35	45		1/176 R
19.9	St Albans	24.30			21.50			26.10			1/176 R
24.6	Harpenden	-			-	pws 15*		-			
30.2	Luton	34.50		¾ L	34.20		¼ L	36.50		2¾ L	
37.3	Harlington	41.25	68		40.40	65/easy		43.10	82		1/200 F
49.8	Bedford	52.00	72	1 E	51.45		1¼ E	52.45	79	¼ E	
59.7	MP 59 ¾ (Sharnbrook)	62.50	44		62.00	45		62.55	41		1/119 R
65	Wellingborough	67.50	70		66.45	80		67.50	75		1/120 F
72	Kettering	74.50			73.05			74.30			
78.5	Desborough North	81.55	52		79.45	55		82.25	45		1/136 R
82.9	Mkt Harborough	86.45	70/easy		84.30	75		87.25	70/easy		1/132 F
89.7	Kibworth North	92.55	50		91.05	45/easy		93.25	50		1/130 R
95.4	Wigston	99.25			97.45			99.35			1/199 F
99.1	Leicester	104.50		1¼ E	102.15 (100 net)		2¾ E	104.20		1¾ E	

The first run with 1021 shows the 'ideal' – hard uphill, modest downhill speeds, keeping time comfortably. The second run with the lighter load ran hard to recover from the severe p-way check and with time well in hand, eased after Market Harborough. The third run started sluggishly but used high speeds after Leagrave to recover time by Bedford and took it easily from Wellingborough onwards, timekeeping assured. In the runs below both Compounds had the 105 minute schedule well in hand, but 1013 worked energetically uphill taking it easy down, whereas 1032 let speed drop uphill but kept steam on down.

		Leicester-St Pancras, c1910					
		1013			**1032**		
		215 tons			**190 tons**		
Miles	Location	Times	Speeds		Times	Speeds	Gradients
0	Leicester	00.00		¾ L	00.00	T	
3.7	Wigston	06.10			06.30		
7.3	Great Glen	-	57½		-		1/199 R
10.2	Kibworth	13.15			13.40		
16	Market Harborough	18.25	77½		19.00	72	1/238 F
20.8	Desborough	24.10	45		25.40	40	1/132 R
27	Kettering	30.05	67*	1 L	31.05	78	1 L	1/118 F
34	Wellingborough	36.25			37.10	70	
39.3	Sharnbrook summit	42.15	46		44.15	38	1/120 R
49.2	Bedford	50.55	75	¼ E	51.55	80	T
57.3	Ampthill	58.50			60.10		1/200 R
66.3	Leagrave	69.10	52		70.35	50	1/200 R
68.8	Luton	71.35		½ E	73.30		½ L
79.2	St Albans	81.10	easy		83.30		
83.8	Radlett	85.35	65		87.25	75	1/176 F
86.5	Elstree	88.20	56½		90.10	52	1/200 R
92	Hendon	93.15	75		95.10	78/sigs	
97.5	Kentish Town	98.40			101.35	sigs	
99	St Pancras	101.47		2½ E	105.05	T	(103¾ net)

No 1032 figured again in a run during 1911, logged right through to Leeds. With a 210 ton load, it took things fairly easily to the first stop at Luton, with 64 at Hendon falling away to 54 at Elstree and 71½ at Radlett. Luton was left just ¼ minute late and got to Leicester in 70 minutes 40 seconds, 1¼ minutes early. No 1032 reached 79mph on the long descent from Leagrave through Flitwick, fell to 48 only at Sharnbrook and 72 on the descent, passing Kettering ¾ minute early. 53 minimum at Desborough was good and 72½ on the descent, slowed to 40 through Market Harborough, was enough. North of Leicester, 1032 quickly attained 74mph by Syston just five miles out, passed Trent Junction a minute early at 20mph and then was hampered by colliery subsidence slacks through to Chesterfield passed at 65mph half a minute early. The 96½ miles to Leeds were completed in 108¾ minutes and arrival was 1¼ minutes early.

Two runs were tabled in the May 1911 *Railway Magazine*, describing the performance of a couple of Compounds on the 196 mile long Leeds-St Pancras non-stop element of the 10.30am from Edinburgh. The first is of particular interest in that it records the work of one of the five prototype Smith Compounds, numbered between 2631 and 2635, with a good load of 240 tons. The other is of the standard Midland Compound, 1030, with the normal 215 ton load, which was badly delayed in the early stages passing through the colliery subsidence infested infrastructure, but came up from Trent Junction to London in 124 minutes, at exactly 60mph average passing Trent at 20mph to the stand at St Pancras.

		Leeds-St Pancras, c1908					
		10.30am Edinburgh-St Pancras					
		2631 – 2635 (Smith Compound)			1030		
Miles	Location	Times	Speeds		Times	Speeds	Gradients
0	Leeds	00.00		5 L	00.00	7 L	
	Woodlesford	-	70		-		1/350 F
11	Normanton	12.45	50*		13.30	7 L	
	Wath	-	*		-	*	Colliery slacks
33.8	Rotherham	38.15	20*	4¼ L	40.15	8¼ L	
41.5	Killamarsh	47.00	55		48.30	58	1/343 R
49.8	Chesterfield	56.30	52		57.45	55	
57.3	Doe Hill	65.30	49/60		67.30	sigs	1/160 R
76	Trent Junction	83.00	20*	1 E	92.30	20*	10½ L
84.3	Loughborough	92.45			101.15		
92	Syston	101.30	55		108.45	65	
96.8	Leicester	107.00		1 L	113.45	9 ¾ L	
100.5	Wigston	112.00			118.45		
107	Kibworth	119.00	55½		126.00	52	1/199 R
112.8	Market Harborough	124.15	70		131.15	72	1/130 F
117.8	Desborough	130.00	48		137.45	40	1/132 R
123.8	Kettering	135.15	75	¼ L	143.30	68	10½ L
130.8	Wellingborough	141.00	73		149.30	65	
139	Sharnbrook	150.00	42		159.00	40	1/120 R
146	Bedford	156.00	80	1 E	164.45	83	9¾ L	1/119 F
154	Ampthill	164.30	45		173.15	50	1/200 R
163	Leagrave	175.30	sigs 30*		184.00	50	1/200 R
165.5	Luton	179.30			186.30	60	
175.8	St Albans	189.45	60		196.30	65	1/176 F
183.3	Elstree	196.00	75/60		203.30	70/60	1/200 R
188.8	Hendon	201.00	75		208.15	78	1/176 F
194.2	Kentish Town	206.45			213.30		
195.8	St Pancras	210.00	(208½ net)	T	216.30	8½ L (210 net)	

	Nottingham-St Pancras, c1911			
	8.25am Nottingham			
	1018			
	190 tons			
Miles	Location	Times	Speeds	Gradients
0	Nottingham	00.00	4½ L	
8.3	Widmerpool	12.30		1/200 R
14.3	Grimston	19.25	52	1/220 R
18.3	Melton Mowbray	23.25	68	1/220 F
24.3	Whissendine	30.20	52	1/260 R
29.5	Oakham	35.35	60	L
33.5	Manton	38.50	74	1/142 F
38.5	Harringsworth	43.10	77½	1/167 F
44	Corby & Weldon	48.35	58	1/200 R
51.5	Kettering	56.05	60	1¾ L
58.5	Wellingborough	62.55	72	
66.7	Sharnbrook	71.15	47	1/120 R
73.8	Bedford	77.05	80½	1/119 F
81.8	Ampthill	86.35	50 strong side gale	1/200 R
90.8	Leagrave	97.50	48	1/200 R
93.2	Luton	100.20		
103.5	St Albans	109.45	65/80½	1/176 F
111	Elstree	115.55	62	1/200 R
116.5	Hendon	119.35	79	1/176 F
122	Kentish Town	125.35	sigs	
<u>123.5</u>	<u>St Pancras</u>	<u>129.30</u>	T	

This run is included as it is fairly typical of competent Compound performance – light load, good uphill, no more than 80mph downhill. The downhill speeds would normally have been taken even more easily in the low or mid-70s, but 1018's crew were recovering from a late start and encountering a strong side wind.

The Midland Class 4 (Compound & Simple) • 175

		Leicester-St Pancras, c1912					
		1012		1039 + 685 (Johnson 4-2-2)			
		195 tons		330 tons			
		6.10pm ex Leicester		Scotch Express (Xmas Eve)			
Miles	Location	Times	Speeds		Times	Speeds		Gradients
0	Leicester	00.00			00.00		30 L	
3.7	Wigston	06.05			06.30			
7.3	Great Glen	-	51½		-	53		1/199 R
10.2	Kibworth	13.25	71½		13.50	75		
16	Market Harborough	19.10	*		19.00	* (severe)		1/238 F
20.8	Desborough	25.55	38		26.00	38		1/132 R
27	Kettering	31.45	72	1¾ L	31.30	72	31½ L	1/118 F
34	Wellingborough	37.50		¾ L	37.20	73	30¼ L	
39.3	Sharnbrook summit	43.25			42.40	45		1/120 R
42.3	Sharnbrook	-	71½		-	71½		1/119 F
49.2	Bedford	52.45	73	¾ L	51.30	73	29½ L	
57.3	Ampthill	60.25	57		59.40	51½		1/200 R
66.3	Leagrave	70.30	49		70.10	47½		1/200 R
68.8	Luton	73.05		T	72.40		29¾ L	
79.2	St Albans	83.00		T	82.05		29 L	
83.8	Radlett	-			-	82		1/176 F
86.5	Elstree	90.15			88.00	64½		1/200 R
92	Hendon	95.10			92.35	79		
	Finchley Road	98.35			95.35	sig stand		
99	St Pancras	103.40		1¼ E	104.20		29¼ L	(100½ net)

The run with 1012 is a perfect example of the ease with which a compound could keep the 105 minute schedule without any undue effort. Nothing higher than 72mph was noted by the recorder and where the highest speeds were often noted (at Radlett and Hendon) the average speeds were barely more than 60mph. The run with 1039 is of interest as it shows a heavier load with the assistance of a Johnson 'Single'. Despite the lateness of the train, speeds were moderate until the last dash for the terminus spoiled by the dead stand outside St Pancras.

Derby-Manchester, c1912
5pm St Pancras-Manchester
1016
162/170 tons

Miles	Location	Times	Speeds		Gradients
0	Derby	00.00			
5.3	Duffield	09.25	sigs		
10.4	Ambergate	16.00	sigs		
17.2	Matlock	25.10			1/177 R
21.6	Rowsley	30.05			
25	Bakewell	35.00			1/102 R
27.3	Longstone	39.00	34½		1/100 R
31.4	Millers Dale	45.30	38		1/100 R
36	Peak Forest	53.20	32½		1/90 R
41.6	Chinley	59.45			1/90 F
44.3	New Mills S. Jcn	62.35	60		1/89 F
53.4	Cheadle Heath	71.45			1/100 F
59.2	Chorlton	77.00			
61.4	Manchester	82.50	(80 net)	1¾ L	

Manchester-Derby, c1912
12.20pm Manchester-St Pancras
1013   1011
229/245 tons   255/270 tons

Miles	Location	Times	Speeds		Times	Speeds		Gradients
0	Manchester	00.00						
4.2	Chorlton	05.55						
8	Cheadle Heath	11.35		1½ E				
		00.00		T	00.00		T	
4	Hazel Grove	08.00			08.40	42½		1/100 R
9.1	New Mills S. Jcn	15.45	45	1¾ L	15.40	49/42½		L/1/100 R
11.8	Chinley	19.50	42		20.00	38		1/90 R
13.7	Chapel-en-le-Frith	22.50	38		23.15	36		1/90 R
17.4	Peak Forest	29.15/31.05	sigs	¼ L/2 L	29.30	34½	½ L	1/90 R

Manchester-Derby, c1912
12.20pm Manchester-St Pancras

| | | 1013 | | | 1011 | | | |
| | | 229/245 tons | | | 255/270 tons | | | |
Miles	Location	Times	Speeds		Times	Speeds		Gradients
22	Monsal Dale	36.55		3 L	34.40	sigs	¾ L	1/90 F/ 1/100 F
28.4	Bakewell	42.40	67		41.00	75		1/100 F
31.8	Rowsley	45.45	68	1¾ L	44.00	68/60*	T	1/100F/1/184 F
36.2	Matlock	50.20			48.15	65		
43	Ambergate	58.10		2 L	56.35	sigs/61	½ L	
48.1	Duffield	64.05	68		62.25	68		
52.6	Nottingham Road	69.00	50		66.40	60		
53.4	Derby	70.35	(67 net)	2½ L	67.30	(66 net)	1½ L	

Manchester-Derby, c1912
12.20pm Manchester-St Pancras

| | | 1026 | | | 1012 | | | |
| | | 214/225 tons | | | 211/230 tons | | | |
Miles	Location	Times	Speeds		Times	Speeds		Gradients
0	Manchester	00.00			00.00			
4.2	Chorlton	07.00	sigs		06.00			
8	Cheadle Heath	13.15		1¼ L	11.15	57½	¾ E	
12	Hazel Grove	19.00	37½		16.45			1/100 R
17.1	New Mills S. Jcn	27.30	34½	1½ L	25.45	34½	¼ E	L/1/100 R
19.8	Chinley	32.30	31½		31.00	31½		1/90 R
21.7	Chapel-en-le-Frith	36.25	29 ½		35.25			1/90 R
25.4	Peak Forest	44.50	25	1¾ L	43.35	26	½ L	1/90 R
30	Monsal Dale	49.45	70	1¾ L	48.35		½ L	1/90 F/ 1/100 F
36.4	Bakewell	55.45	72		55.00			1/100 F
39.8	Rowsley	58.45	sigs	¾ L	58.10	68	¼ L	1/100F/1/184 F
44.2	Matlock	64.30	65		62.55			
51	Ambergate	72.50		1¾ L	70.15		¾ E	
56.1	Duffield	78.50	68		76.35			
60.6	Nottingham Road	83.05	sigs		80.45			
61.4	Derby	85.00	(81½ net)	2 L	82.30	(82½ net)	½ E	

These five runs between Derby and Manchester are of interest as they show typical performances, mostly with slightly heavier loads, over the steep gradients to Peak Forest. At their best, the compounds were capable of climbing the long 1 in 100s with 250 tons in the mid-30s, though the last two logs are not in the same league as the previous two. No 1012 had the logic of running to time so no extra effort was needed, but 1026 failed to recover from the signal checks earlier in the run and only 25mph at Peak Forest, just sufficient if running on time, was below par.

The Midland started a policy of grouping engines in the same numerical sequence together – possibly to identify from quick observation an engine's home base. This rationalisation led to the following allocations in 1914:

Leeds:	1000-1010
Manchester:	1011-1021
Nottingham:	1022-1030
Kentish Town:	1031-1044

This allocation held through most of the First World War though two of the Leeds engines (1002 and 1010) were transferred to Carlisle in 1917, still however working over the same route. Then, between 1917 and 1920 Nottingham lost its entire fleet of compounds which were dispersed, the 1920 allocation being:

Leeds:	1000, 1001, 1003-1009, 1025, 1027, 1028
Carlisle:	1002, 1010, 1023
Manchester:	1011-1022, 1026
Kentish Town:	1024, 1029-1044

After the Grouping in December 1923, the LMS authorities tested the two Midland class 4 4-4-0s, 1008 still saturated and the superheated 998, and the former LNWR 'Prince of Wales' 4-6-0 388. The trials were conducted between Leeds and Carlisle in both directions and the loads were substantially more than the Midland engines normally hauled over that route, varying from 300-350 tons tare. Time was kept on nearly all the turns, apart from a couple of trips with 998 on 350 tons when it failed to maintain steam production on the upper part of the bank approaching Blea Moor and lost 5-7 minutes. No 1008, which had always had a good reputation, was the most consistent hill climber. Southbound with 320 tons gross, 1008 passed Appleby on time (39 minutes exactly for the initial 30.8 miles), achieved almost 70mph in the dip before Ormside, fell to 42½ on the first stretch of

Kentish Town's Deeley compound 1013 (built as 1008) in December 1905, on a down express at Elstree, c1910. (E.M. Johnson Collection)

The Midland Class 4 (Compound & Simple) • 179

Kentish Town's 1006 on a down fast train at Thurmaston, just north of Leicester, c1910.
(MLS Collection)

Kentish Town's 1012 passing Wigston with a Manchester-St Pancras express, c1910.
(G.M. Shoults/MLS Collection)

No 1017 leaving Manchester Central with an express for St Pancras, 30 June 1911.
(MLS Collection)

The rebuilt prototype 2631, with new frames and superheated boiler, on test on an up express near Mill Hill, 1914.
(E. Pouteau/MLS Collection)

Carlisle's 1023 with an up express near Leeds, c1922. No 1023 would be superheated in November 1923. (F. Moore/ MLS Collection)

1 in 100, recovered to 50 on the easing after Crosby Garrett, fell to 32 before Birkett Tunnel, accelerated to 41 on the short stretch of 1 in 330 at Mallerstang and cleared the summit at Ais Gill at 32mph, gaining a full five minutes on the schedule over this difficult section. The following day, 18 December, 1008 had 370 tons gross and got to Appleby even faster than the previous day, and its speeds on the key points of the climb were 65 at Ormside, 37/47 at Crosby Garrett, 27 at Birkett Tunnel, 37½ at Mallerstang and 27 at Ais Gill, now 3½ minutes early, having gained 2¼ minutes on the train's schedule between Appleby and Ais Gill with a load over 100 tons heavier than normally provided on this service. No 1008 was producing 1,025 drawbar horsepower on the second run on the 1 in 132 before Langwathby surmounted at 56 ½mph. Coal consumption was 42.4lbs per mile (3.93 per dhp) on the first run and 46.2 (3.83 per dhp) on the second. Neither of the other engines tested achieved the same power output and the coal consumption of the LNWR engine was much heavier. In the down direction with 305 tons gross 1008 sustained 35-40mph over most of the climb falling to 33mph on the final couple of miles between Ribblehead and Blea Moor. For a full table of the key logs and comparison with class 990, No 998, see pages 201-2.

The data from the tests justified the preference of the former Midland management team, then holding the key roles in the new company, for the Midland compound to be chosen over the LNWR engines for the future LMS main line express power, 195 LMS Fowler compounds being built at Derby throughout 1924 and 1925 at more than one a week. The influx of the new compounds caused some movements among the Midland engines. Carlisle lost two of its three, 1010 going to Leeds and 1023 to Kentish Town. Manchester lost most of its allocation to Leeds and Kentish Town though it gained 1014 from Leeds. Some trials of compounds on the former LNWR routes took place and records exist of a couple of runs with 1033 between Birmingham and London. With only moderate loads scheduled time was kept comfortably without the need for anything over 75mph.

		Birmingham New Street-Willesden Junction/Euston, 1924				
		1033		1033		
		209/220 tons		189/200 tons		
Miles	Location	Times	Speeds	Times	Speeds	Gradients
0	Birmingham New St	00.00		00.00		1/58 F
1.9	Adderley Park	03.48	45	03.45	50	L
3.8	Stechford	05.57	60	05.50	61	1/660 F
6.5	Marston Green	08.25	70	08.25	66	
10.1	Hampton-in-Arden	11.28	70	11.35	70	L
13.4	Berkswell	14.25	63	14.40	61	1/330 R
15.4	Tile Hill	16.10		16.30	72	1/330 F
18.9	Coventry	19.08	75	19.35	60*	1/330 F
23.8	Brandon	23.05	76	24.15	70	L
29.7	Rugby No.7	28.40	44*	30.00	45*	
30.3	Rugby	29.35	40*	30.55	35*	
37.6	Welton	38.30	61	39.50		1/370 R
43.2	Weedon	43.20	73	44.45	74	1/350 F
50.1	Blisworth	49.25	64	51.00		L
53	Roade	52.05	62	53.55	60	1/320 R
58.1	Castlethorpe	56.30	75	58.20	75	1/326 F
60.5	Wolverton	58.30	69	60.25		L
66.2	Bletchley	63.38	64	65.55	64	1/440 R
72.7	Leighton Buzzard	69.42	62	72.25	60	
76.8	Cheddington	73.37	58	76.30	55	1/333 R
81.2	Tring	78.20	52	81.25	54	1/333 R
84.9	Berkhamsted	82.05	63	83.05	68	1/335 F
	Hemel Hempstead	-	69	-	72	1/335 F
91.9	King's Langley	88.08	70	91.05	70	
95.4	Watford	91.18	60 eased	94.25	55 eased	
96.9	Bushey	92.46	58	96.10	easy	
98.6	Hatch End	95.35	56	99.55		1/339 F
101.5	Harrow & Wealdstone	97.25	62	102.00		
104.8	Wembley	100.46	52	105.40		
107.5	Willesden Junction	104.30	5½ E	109.35		
112	Euston			119.00	¼ E	

The Midland Class 4 (Compound & Simple) • 183

Manchester's 1012, superheated in May 1922, on an up express near Duffield between Ambergate and Derby, c1925. (F. Moore/MLS Collection)

Leeds Holbeck's 1027 waiting to depart from St Pancras with a down express as the first LMS-built compound, 1045, stands ready in the next platform, 1930.
(G.A. Coltas/MLS Collection)

Superheated 1044 on a Derby-Manchester express eases after passing through Dove Holes Tunnel, c1930. (MLS Collection)

By the 1930s, the compounds had lost their main express turns initially to ex LNWR Claughtons, then to 'Patriots', Stanier 'Black 5s' and 'Jubilees'. The Derby-Bristol route had been upgraded and the allocation of the 45 Midland compounds in the early 1930s was:

Leeds:	1008-1013,
Manchester:	1017-1022
Kentish Town:	1023-1027, 1032-1043
Leicester:	1003, 1004
Sheffield:	1005-1007, 1016
Derby:	1014, 1015, 1044
Bristol:	1028-1031
Gloucester:	1000-1002

Then, in the mid-1930s, 1023-1027 augmented the fleet at Bristol, 1038 came to Bedford and 1040-1043 were moved from Kentish Town to Leicester and the compounds began to take on the semi-fast and stopping services between London, Bedford and Leicester. Nos 1006, 1008 and 1036 went to Carnforth for the Leeds-Hellifield-Lancaster/Carnforth services. No 1008 then wandered all over the system between 1936 and 1939 – Workington, Longsight, Rugby and Willesden, returning to Leicester at the beginning of the Second World War. At the end of the war, the Midland compounds were based at Leeds, Manningham, Hellifield, Lancaster, Sheffield, Manchester, Derby, Nottingham, Leicester, Bedford, Saltley, Gloucester and Bristol. Fifteen (one third of the class) were based at Leicester and Bedford for local and semi-fast services at the southern end of the Midland main line. Some of the suburban commuter services would load to 350 tons gross – a more strenuous life than in their heyday on the lightweight Midland expresses before the First World War.

Bristol's 1028 at Cheltenham Junction with a Bristol-York express, 18 March 1933.
(E.R. Morten/MLS Collection)

No 1029 of Bristol awaiting departure from Derby with a train for Bristol, 30 April 1933.
(E.R. Morten/MLS Collection)

Kentish Town's 1034 at St Albans on a down stopping train to Bedford and Leicester, c1935. (Photomatic/MLS Collection)

No 1024 at Chapel-en-le-Frith with a Manchester-Derby stopping train, 18 May 1937. (MLS Collection)

The Midland Class 4 (Compound & Simple) • 187

Manchester's 1020 at Romiley with a Manchester-Sheffield stopping train, 4 September 1937.
(MLS Collection)

No 1009 approaches the summit of the 1 in 90 at Great Rocks Junction, Peak Forest, with a Derby-Manchester train, 29 September 1938.
(E.R. Morten/MLS Collection)

No 1003 (the former 2634) performing humble unfitted freight duties at Hartford on the West Coast Main Line in wartime conditions, 4 May 1940. (G. Harrop/MLS Collection)

All were taken into BR ownership in January 1948, but 1029 was stored at Derby Works pending a decision in January. It was among the first to be withdrawn in the summer of 1948, with 1002, 1018, 1026 and 1033, coincidentally taken out of traffic in the same month as the very last ex-LNWR 'George V' 4-4-0, 25350, formerly named *India*. By 1950, just twenty-eight Midland compounds remained at the following locations:

Leeds:	41040
Manningham:	41004
Lancaster:	41005
Sheffield:	41014, 41016, 41021, 41037
Nottingham:	41015, 41019, 41032
Derby:	41000, 41003, 41023, 41043
Leicester:	41006, 41011, 41041
Kettering:	41012
Bedford:	41007, 41009, 41038, 41044
Kentish Town:	41020
Saltley:	41035
Gloucester:	41001, 41025
Bristol:	41028, 41030

These were withdrawn between 1950 and 1952, the last survivors being 41004 withdrawn in February 1952 from Manningham, 41032 in March from Nottingham, 41014 from Derby, 41007 and 41009 from Bedford, 41035 from Bourneville and 41040 from Leeds in May, 41038 from Bedford in August, 41021 from

The Midland Class 4 (Compound & Simple) • 189

Two photos of 41003 taken on the same day on a Derby-Manchester stopping train at New Mills Junction and at Gow Hole on the return working with four of the five coaches of the outward journey, 28 August 1948. Note that although it has recently been repainted in BR lined mixed traffic livery, it has not been yet provided with a renumbered smokebox door numberplate.
(MLS Collection)

Sheffield's 41016 departing from Chinley with the 3.16pm stopping train to Manchester, 4 September 1949.
(J.D. Darby/MLS Collection)

No 41007 of Bedford shed on a St Pancras-Bedford stopping train approaching Luton station, 12 June 1950. No 41007 remained a Bedford engine on such work until its withdrawal in May 1952.
(H.C. Casserley/MLS Collection)

A filthy 41014, still showing LMS on its tender, with a Manchester-Derby stopping train, 4 August 1951. No 41014 was withdrawn in May 1952. (MLS Collection)

Nottingham, 41028 from Bristol and 41044 from Bedford in October, and finally 41025 from Gloucester in January 1953.

Preservation
1000
No 41000 was withdrawn in September 1951 and stored at Derby pending a decision on preservation. It moved to Crewe Works in 1953 and stayed there until 1959 when it returned to Derby for restoration to its post superheating 1914 condition. Frame repairs were carried out and a 1947 built boiler replaced the withdrawn one. Its tender was replaced by a 1914 Deeley tender off an S&DR 2-8-0 (53805). It has the tall Deeley chimney and the Midland crest on the cabside and large 1000 numerals on the tender. Its thorough restoration was to running standard and between 1959 and 1962 it had occasional use of special trains, mainly railtours, based at Derby. One of the most impressive of these was on Sunday 4 September 1960, when

it hauled a nine coach *Cumbrian Railtour* from Leeds to Carnforth, handing over to unrebuilt Patriot 45503 to Ravenglass for a number of specials on the narrow gauge line, then to Workington where a couple of Ivatt class 2 moguls took the train to Penrith, where 1000 reappeared to complete the run to Carlisle and then ran south via the Settle and Carlisle back to Leeds. The section from Carlisle to Hellifield was timed by Mr D. Twibell and I table the log below.

Carlisle-Hellifield, The Cumbrian Railtour
1000
9 chs, 288/310 tons

Miles	Location	Times	Speeds	Gradients
0	Carlisle	00.00		
2.7	Scotby	05.49	26½	1/132 R
8	MP 300 (Cotehill)	-	40/33	1/132 R
9.8	Armathwaite	17.33	53/46	1/132 F
15.4	Lazonby	23.44	60/ pws 17*	
23.3	Culgaith	36.48	46	
	Long Marston	-	51/38	L, 1/120 R
30.8	Appleby	46.37	44½	
33	MP 275 (Ormside)	49.35	56½	1/176 F
	Griseburn	-	30½	1/100 R
	Crosby Garrett	-	40/45	1/200 R, L
41.8	Kirkby Stephen	62.28	33/36/30	1/100 R
	Mallerstang	68.31	39	
48.3	Ais Gill	74.35	31	1/100 R
	Garsdale	-	1* slack	
54.6	Dent	86.33	57	
59.5	Blea Moor	91.51	51/60	
65.5	Horton-in-Ribblesdale	96.38	80	1/100 F
	Stainforth Sidings	99.46	84	1/100 F
71.5	Settle	101.01	87	1/100 F
	Settle Junction	102.29	78/sigs	
<u>76.8</u>	<u>Hellifield</u>	<u>112.19</u>	(102 net)	

The Midland Class 4 (Compound & Simple) • 193

The restored 1000 at Doncaster after working a special train from Derby, 30 August 1959. (J. Lord/MLS Collection)

No 1000 at Trafford Park shed, Manchester, 1 April 1962. (MLS Collection)

It was placed in the Clapham railway museum in November 1962 and transferred to the NRM at York in 1975 when Clapham closed. It was put into working order again in 1980 when it ran a number of railtours, including a run over the Settle & Carlisle in 1983 piloting 'Jubilee' 5690 *Leander*. I had one run on 3 May 1980 when it piloted 4771 *Green Arrow* on the Leeds-Carnforth leg of the *Mancunian* railtour which had started from Piccadilly behind English Electric diesel 40020, traversed the Woodhead route behind electrics 76011 and 76021 (ex 26011/21), then reverted to type 4 diesel haulage with 40083. The special was restricted to 60mph on the steam legs and I give highlights of the run with 1000 below. The train left Leeds 43 minutes late (after late departure from Manchester following a diesel failure) and recouped fifteen minutes to Hellifield over the gradually rising gradients just touching 60mph after Keighley and 65 before Skipton. The minimum on the 1 in 132 to Bell Busk was 38mph. Hellifield was left 16½ minutes late and the log of the running to Carnforth is shown below.

It returned after these runs to static exhibit in the NRM's Great Hall at York, although currently (2020) it is on loan to the Barrow Hill roundhouse display in Derbyshire.

Leeds City-Carnforth, 3 May 1980
1Z77 The Mancunian
4771 + 1000
11 chs, 385/420 tons

Miles	Location	Times	Speeds	Gradients
0	Hellifield	00.00		16½ L
2.5	Settle Junction	05.13	54/20*	
4.8	Giggleswick	08.31	38	1/100 R
6	MP 237 ¾	-	41	1/100 R
10.3	Clapham Junction	16.00	65/25*	
14.5	Bentham	20.43	71/69	1/100 F & 1/180 F
17.9	Wennington Junction	24.05	72/40*	16½ L 1/141 F

Despite a signal stand outside Carnforth, the engine exchange point there was reached just eleven minutes late. The 60mph limit was (nearly) kept over the last twelve minutes where the preserved pair were replaced by 4498 *Sir Nigel Gresley* for the return to Manchester Piccadilly via Hellifield, Blackburn and Bolton.

The Class 4 '990' Simple
Design & construction
As early as 1906, Deeley had produced a drawing of a 4-4-0 very similar to the compound, except with two cylinders and 'simple' rather than compound valve and motion arrangement. Then, a few weeks later, a second drawing was produced with 6ft 6in coupled wheels instead of 7ft and 19in diameter cylinders instead of 19½in. A single locomotive was authorised, presumably to compare with the compound, and No 999 was built and entered traffic in March 1907. Its dimensions were:

Cylinder
  diameter   19in x 26in
Walschaerts valve gear
  (Deeley 'scissors' version)
Coupled wheel
  diameter   6ft 6½in
Bogie wheel
  diameter   3ft 3½in
Boiler pressure   200lbs psi
Grate area   28.4sq ft
Heating surface   1,557.4sq ft
Axleload   19¾ tons
Weight
  – Engine   58 tons 10 cwt
  – Tender   45 tons 18 cwt
  – Total   104 tons 8 cwt
Water capacity   3,500 gallons
Coal capacity   7 tons
Tractive effort   23,662lbs

Extensive tests took place in 1907 and 1908, covering power output and coal consumption and as a result nine more numbered 990-998 were constructed between May and October 1909. As a result, they were known as the '990' rather than '999' class. Deeley had considered the enlargement of his 4-4-0 Compound as a 4-6-0 and in 1907 Derby drawing office had produced an outline drawing of a 4-cylinder compound 4-6-0 (see Appendix, page 341), but this was rejected as unnecessary

The Midland Class 4 (Compound & Simple) • 195

The prototype, 999, as built in March 1907, in Works grey for photographic purposes.
(F. Moore/MLS Collection)

No 995, built in July 1909, at Leicester station, 9 April 1910. Note the saddle under the slightly extended smokebox which differentiated them from the early Compounds.
(W. Bradshaw/MLS Collection)

No 992 at Derby after rebuilding with a superheated boiler in August 1912. Note the Davies and Metcalfe exhaust steam injector under the extended smokebox which was fitted at the same time (995 and 993 had been equipped earlier).
(MLS Collection)

and the building of the '990' order and more 4-4-0 Compounds were pursued instead.

The decision to fit some Midland locomotives with superheaters was explored in 1909, and it was thought that a Schmidt superheater should be provided to a simple rather than compound engine. The estimated cost per engine was £256 including a royalty of £50 to the Schmidt company. The new class 990s were chosen to be the vanguard and boiler type G9AS was fitted to 998 in May 1910. The superheater surface was 360sq ft, giving a total heating surface of 1,681sq ft. No 999's boiler was rebuilt to the same specification and the rebuilding was complete by May 1911. No 998's original boiler was then modified for superheating in the same way as for 999 and was fitted to 995 in March 1912.

Tests had been undertaken with the superheated engine 998 in October 1910, comparing it with the still saturated 999 and although details of the trials are sketchy, the authorities drew the conclusion that the superheated engine had an advantage in coal consumed per indicated horsepower over an hour's running time of nearly 40 per cent. Whatever the margins of error, this was more than enough to persuade Fowler and the Board to authorise the fitting of superheated boilers not just to the rest of the '990' class, but to its entire 4-4-0 fleet, with a significant programme of conversions in the immediate years before the First World War.

The remaining members of the class were therefore rebuilt between July 1912 and January 1914. Boiler pressure on the superheated engines was reduced to 180lbs psi as a means of reducing costs, offset by equipping the engines with larger 20½in x 26in cylinders. The weight increased by 1¾ tons and tractive effort was 20,043lbs.

The average time between Works overhauls for the class when concentrated on the Settle & Carlisle line was nineteen months when the engines had accumulated around 50,000 miles. All ten of the locomotives were absorbed into the LMS fleet in 1923, but they were a small non-standard fleet and although they were involved in 1924 tests with a compound and LNWR 4-6-0,

The Midland Class 4 (Compound & Simple) • 197

The former 993, renumbered 803 in May 1926, at Derby. The exhaust steam injector is very obvious here – only 992, 993 and 995 had received this equipment. No 803 was withdrawn in January 1928. (Real Photographs/MLS Collection)

No 807 (ex 997), superheated in July 1913, renumbered in April 1927, and withdrawn in December 1927. The cylinder by-pass valve is seen behind the smokebox to prevent knocking of the connecting rod big and little ends when running with the regulator shut. (W.L. Good/MLS Collection)

the test engine, 998, did not perform as well as the others. As the number of LMS Compounds rapidly increased, the 990s lost their roles and 990 itself was withdrawn as early as May 1925 and 994 followed in 1926. As the construction of Compounds numbered in the 900 series grew, it was decided that the 990s should be renumbered leaving space for the latter engines and those remaining were numbered 801, 803, and 805-809, as 990, 992 and 994 had been withdrawn before the renumbering. All had gone by 1928, 808 (998) and 809 (999) being the last withdrawn in December that year. Some of the boilers were retained and used on Compounds during regular overhauls.

A number of commentators have expressed the view that had the lessons of Churchward's long lap valves and long travel been applied to the '990' class, the advantages shown by the Compounds would have been more than offset and compounding would have finished with the 45 Midland engines.

Operations

999 was based at Derby immediately after construction while it was subjected to running in and a series of tests, which included some express running between Derby and St Pancras. A few minor problems were experienced including piston valves blowing and new valves were fitted which solved the problem and cut coal consumption. When the other nine engines were built in 1909, the ten were then allocated as follows:

Kentish Town:	990-992
Manchester:	993-995
Leeds:	996-998
Derby:	999

Tests took place between 990, 992 and 996 in comparison with Compounds 1010 and 1034 in 1910. The trials took place between London and Leeds with trains varying in weight from 300 to 350 tons, well in excess of those allowed for single engines without pilot assistance then. The superheated 990 was more economical than the saturated Compound but after tests in 1913 with the first superheated compound (1040) that showed significant improvement over the 990 as well as its saturated sisters.

A run recorded in the April 1912 *Railway Magazine* makes an interesting comparison with similar logs of the Compounds. Although overall times are very similar to a log with Compound 1032 (on page 172) it is clear that the Compound had the edge uphill and the crew drove 990 harder downhill. This was particularly noticeable between Luton and Leicester.

		St Pancras-Leicester-Leeds, c1911		
		990		
		180 tons		
Miles	**Location**	**Times**	**Speeds**	**Gradients**
0	St Pancras	00.00	T	
1.5	Kentish Town	03.40	¼ E	1/178 R
6.9	Hendon	10.35	62½ ½ L	1/200 F
12.4	Elstree	16.35		1/176 R
15.2	Radlett	19.20	72½	1/200 F
19.9	St Albans	22.35		1/176 R
24.6	Harpenden	-	65	
30.2	Luton	34.15	¾ E	
0		00.00	½ E	
7.5	Harlington	09.45	83½	1/200 F
19.7	Bedford	19.50	74 1¼ E	

The Midland Class 4 (Compound & Simple) • 199

		St Pancras-Leicester-Leeds, c1911			
		990			
		180 tons			
Miles	Location	Times	Speeds		Gradients
29.5	MP 59 ¾ (Sharnbrook)	29.35	41½		1/119 R
34.8	Wellingborough	34.30	76		1/120 F
41.8	Kettering	41.05		1½ E	
48.3	Desborough North	49.20	38/71½		1/136 R
52.7	Mkt Harborough	54.20	50*		1/132 F
56.1	East Langton	-	67		
59.5	Kibworth North	60.55	48		1/130 R
65.2	Wigston	67.55	50*		1/199 F
<u>68.9</u>	Leicester	<u>72.35</u>		T	
0		00.00		T	
4.8	Syston	06.50			
12.5	Loughborough	13.55	72½		1/508 F
20.7	Trent Junction	21.40	30*	¼ E	
39.5	Doe Hill	46.10	sigs		1/230 R Colliery slacks
47	Chesterfield	55.20	30*	1¼ L	
55.5	Killamarsh	63.55	66		1/326 F
63	Rotherham	71.05	sigs 20*	T	Colliery slacks
<u>96.5</u>	<u>Leeds</u>	<u>111.35</u>		<u>1½ L</u>	

Nos 993-995 were transferred from Manchester to Carlisle around 1912 and joined Leeds' 996-998, their 6ft 6in coupled wheels being considered more appropriate for the heavily graded line. The *Railway Magazine* of September 1913 recorded a number of runs over the Settle and Carlisle line with class 2 4-4-0s, Compounds and the 990s. I give below four of the runs with the 990s.

		Hellifield-Carlisle, c1912							
		996 +465		995		995		993	
		285 tons		255 tons		175 tons		245 tons	
Miles	Location	Times	Speeds	Times	Speeds	Times	Speeds	Times	Speeds
0	Hellifield	00.00		00.00		00.00		00.00	
2.3	Settle Jcn	04.25	½ L	04.25	½ L	04.30	½ L	05.00	1 L
5.2	Settle	06.25		06.20		06.30		07.05	
11.3	Horton	14.30	48	14.25	47	14.45	48	16.00	45
17.3	Blea Moor	22.45	43½ 2¼ E	23.30	40½ 1½ E	23.35	41	25.20	38

			Hellifield-Carlisle, c1912							
			996 +465		995		995		993	
			285 tons		255 tons		175 tons		245 tons	
Miles	Location	Times	Speeds		Times	Speeds	Times	Speeds	Times	Speeds
22.2	Dent	28.30	60		29.50	easy	29.30	70½	31.35	
25.4	Hawes Jcn	31.35			33.15	49	32.40	57½	34.55	
28.5	Ais Gill	34.35/35.45	det 465	1¼ E	36.20	¾ E	35.35	2½ E	38.05	1 L
35.3	K'by Stephen	44.00			43.15		43.50	sigs	44.10	
43.6	Ormside	51.30	71½		51.40	sigs	51.50		51.45	76½
46	Appleby	54.00		1 E	55.00	T	54.40	1¼ E	54.45	1¼ E
							55.40	1¼ E	55.30	1½ E
53.5	Culgaith	60.55			61.45		64.30		64.40	
61.4	Lazonby	68.10	71½		67.55	83½	71.05	77½	71.55	70
66.8	Armathwaite	73.50			72.45		76.06		77.35	
74	Scotby	89.45	sigs stand (2)		78.50	72	82.25		84.20	
76.8	Carlisle	89.45		3¾ L	83.25	2½ E	87.00	2 E	88.05	T
		(85¼ net)			(81½ net)		(85½ net)		(88 net)	

No 996 on the first run was still unsuperheated and was assisted to Ais Gill by Leeds based '2203' class 465 which had been equipped with a G7 Belpaire boiler in 1910. No 995 in the second run had been superheated and was the best of the runs maintaining over 40mph on the long 1 in 100 with a substantial load. The third is more typical of normal 990 running on this route with the usual light load and well within schedule. The last run is again with a bigger load but time was kept, with very easy running after Appleby.

No 993 was involved in an accident at Ais Gill on 2 September 1913 when it and a pilot engine on an express collided with two light engines due to a signalman's oversight. There were sixteen fatalities and thirty-eight were injured. By 1914, the whole class was based at Carlisle and they spent the rest of their comparatively short careers there, apart from one oddity when 995 was tried on the Somerset & Dorset Railway in 1925 on the assumption that it might prove more effective than the class 2 4-4-0s, but it was not considered successful and returned to Carlisle.

No 992, still unsuperheated, departing from Carlisle Citadel station with a heavy Glasgow-St Pancras express, c1910. It was superheated in August 1912. (MLS Collection)

No 992 again on the Settle & Carlisle line, near Armathwaite, c1911. (MLS Collection)

At the end of 1923 and the beginning of 1924, the LMS conducted engine trials over the Settle & Carlisle line, comparing an LNWR 'Prince of Wales' 4-6-0 with Midland Compound 1008 and Midland simple 4-4-0, 998.

I have referred to this earlier in the chapter on the Midland Compounds (page 181) and again in the next chapter, but give below the full logs of some of the three engines' northbound performance and 998 and 1008 southbound.

Whilst the performances look at first sight similar, 1008's coal consumption was significantly lower than that of 388, and 998 appeared winded on the last few miles before Blea Moor but recovered with brisk running downhill and after Appleby.

Leeds-Carlisle locomotive trials, December 1923
1.30pm St Pancras-Glasgow (4.7pm Leeds)

Miles	Location	388 4-6-0 'Prince of Wales' 291/300 tons			1008 294/305 tons			998 299/310 tons			Gradients
		Times	Speeds		Times	Speeds		Times	Speeds		
0	Leeds	00.00			00.00			00.00			
0.8	Holbeck	02.26			02.23			02.57			
3.1	Kirkstall	06.20	50/47		06.26	49½/46		06.36	50/47		1/264 R
10.9	Bingley	15.59	32*	1 L	16.08	26*	1¼ L	16.28	27*	1½ L	1/290 R
17	Keighley	23.49	51/47	1¾ L	24.05	53/51	2 L	24.33	52/48	2½ L	1/218/244 R
26.2	Skipton	33.39	59/37*	¾ L	33.41	62/39*	¾ L	34.11	60/36*	1¼ L	L
29.9	Gargrave	39.15	44/39		38.55	46/42½		39.28	47/43		1/165 R
32.8	Bell Busk	43.29	43		42.45	47		43.19	46½		1/131 R
36.2	Hellifield	48.09		1 L	47.11		¼ L	47.38		¾ L	
39.5	Settle Jcn	51.30	63½	½ L	50.28	66	½ E	50.38	73	¼ E	1/214/181 F
41.4	Settle	53.40			52.25			52.33			1/100 R
45.8	Helwith Bridge	61.03	34/41		59.06	35½/41		59.18	32/38		1/100 R/L
47.5	Horton	63.32	36/38		61.17	38/39½		61.41	32/35		1/100 R
52.2	Ribblehead	71.21	34		69.00	34		70.15	30½		1/100 R
53.5	Blea Moor	73.29	38	½ L	71.12	37½	1¾ E	72.31	33½		1/100 R
54.5	Blea Moor Tnl	75.11	34½		72.59	33½		74.31	29		1/100 R
58.4	Dent	79.55	55/51		77.45	54/51		79.29	55/53		1/440 F 1/264 R
61.6	Hawes Junction	83.17	64		81.07	65½		82.48	67		L
64.7	Ais Gill	86.25	53½	½ L	84.18	52½	1¾ E	85.48	56½	¼ E	1/165/323 R
79.8	Ormside	100.29	71½		97.37	75		99.55	74		1/100 F
82.2	Appleby	102.50	57	1¼ E	99.56	55	4 E	102.40	45	1¼ E	1/176 R
89.2	Culgaith	-	69		-	66		-	76½		1/120/440 F
97.6	Lazonby	117.19	50		115.00	50		116.08	58		1/264/165 R
	Armathwaite	-	66		-	64		-	72		1/220 F
99.5	Low House	-	55½		-	53½		-	61½		1/132 R
110.3	Scotby	129.42	69		128.00	63		127.29	75		1/132 F
113	Carlisle	133.55		1 E	132.11		2¾ E	132.27		2½ E	

Carlisle-Leeds locomotive trials, December 1923

Miles	Location	998 310/320 tons Times	Speeds		1008 396/320 tons Times	Speeds		Gradients
0	Carlisle	00.00			00.00			
2.7	Scotby	06.19	32/40		06.23	32/41½		1/132 R
8.4	Low House	15.52	36/53		15.42	34/50		1/32 R / 1/132 F
15.4	Lazonby	23.05	70½	2 L	23.19	66	2¼ L	1/165 F
19.8	Langwathby	27.01	56½		27.36	51		1/132 R
23.4	Culgaith	30.33	61		30.53	59½		L
30.4	Appleby	38.29	61/54	½ E	39.00	64/55	T	1/120 R
33.2	Ormside	41.01	65		41.19	69½		1/176 F
36.5	Griseburn	-	38½		-	42½		1/100 R
38.3	Crosby Garrett	47.35	48		47.30	50		1/300 R
41.5	Kirkby Stephen	52.05	34		51.46	38		1/100 R
44.8	Mallerstang	58.19	30/38		57.23	32/41		1/100 R / 1/330 R
48.3	Ais Gill	64.27	28	3½ E	63.03	32	5 E	1/100 R
51.4	Hawes Jcn	68.09	61		66.47	59½		1/165 F
	Dent	-	55		-	53		1/364 R/L
59.4	Blea Moor	76.29	62/56	3 ½ E	75.20	61/55	4½ E	1/440 R
73.5	Settle Junction	88.53	72	4 E	86.58	78	6 E	1/100 F
76.8	Hellifield	92.27		4½ E	90.13	22*	6¾ E	
	Bell Busk	-	43		-			1/214 R
86.8	Skipton	103.14	66/30*	5¾ E	102.48	63/40*	6¼ E	1/165 F
96	Keighley	114.22	20*	4½ E	113.29	16*	5½ E	
102.1	Bingley Jcn	-	sigs/66		-	sigs/63		1/330 F
113	Leeds	140.47		2¼ E	140.24		2½ E	
		(134¾ net)			(131¾ net)			

No 998 did well on this run, though 1008, after taking it relatively easily to Appleby, then showed its superiority on the climb to Ais Gill. Both running early on a schedule designed for 200 tons or less, not 300 tons, restrained speed on the descent from Blea Moor and delays at the end of the journey were inevitable. Subsequent tests with 370 tons widened the gap in performance between the 990 and Compound, the simple engine finding the heavier load more of a struggle.

Once the Compounds were superheated and many drafted to Leeds and Carlisle, and the LMS Compounds came on the scene also, the 990s were relegated to stopping services to Appleby and other local routes, the last run recorded being 809 on an Appleby local on 7 November 1928. No 808 and 809 were withdrawn in December and the class was extinct.

The Midland Class 4 (Compound & Simple) • 203

No 992 seems more camera friendly than its sisters. Here is 992 again, now superheated, departing from Carlisle towards Leeds with a Scotch express, c1920. (MLS Collection)

No 997 arriving at Carlisle Citadel station with a St Pancras-Glasgow St Enoch express, February 1922. (F.E. Mackay/MLS Collection)

Chapter 5

THE LMS OPTIONS

The London & North Western Railway had amalgamated with the Lancashire and Yorkshire Railway in 1922, just a year before the Grouping. Charles Bowen-Cooke, Chief Mechanical Engineer of the LNWR, had died in 1920 and had been replaced by his assistant, H.P.M. Beames, but George Hughes, Chief Mechanical Engineer of the L&Y, had assumed control as the senior mechanical engineer of the two companies and the same logic held sway when the combined company was amalgamated with the Midland and Scottish companies into the London Midland & Scottish Railway in 1923, when he became its first CME.

The LMS had inherited some 4-4-0 and 4-6-0 classes for express passenger train work and a number had been designed and constructed in relatively recent times and were in charge of the main passenger services on their respective companies' routes. George Hughes, as an L&Y man, looked first at his own designs, particularly his 1919 rebuild of the earlier 4-cylinder 4-6-0s of 1908. Fifteen of the twenty of these 6ft 3in coupled wheel engines had been rebuilt and Horwich was in the middle of building fifty-five new engines to this design, thirty-five of which had been completed before the Grouping. The remaining twenty already authorised were completed by the end of 1924. They weighed 77 tons, were superheated, had Walschaerts valve gear. At the time (and until the building of the first Gresley pacific in 1922 and the Collett Castle in 1923) they were Britain's most powerful express passenger locomotive, with a tractive effort of 28,880lbs though their performance on the road and coal consumption did not match that of Churchward's 'Saints' or 'Stars'. In 1921, they were trialled on the Crewe-Carlisle section against an LNWR Claughton, resulting in the decision to build more of the L&Y rather than the LNWR class. However, E.S. Cox, in a paper in 1946, referred to them as 'poor steamers, with heavy coal consumption and not outstanding reliability'. The last one (50455) was withdrawn in 1951.

When decisions had to be made about future locomotive construction policy after Henry Fowler replaced Hughes in 1925, the heavy coal consumption of the Hughes 4-6-0s counted against them. The LNWR Claughtons had already been tested and found wanting in 1921. No 130 had been built by the LNWR between 1913 and 1921 to the design of Charles Bowen-Cooke, 4-cylinder engines with 6ft 9in coupled wheels, superheated and with Walschaerts valve gear. Boiler pressure was 175lbs psi, and tractive effort 27,072lbs, though twenty were later (1928) rebuilt with larger 200lbs psi boilers. Nearly all had been withdrawn by 1937, some had been reconstituted into the LMS 'Patriot' class, and the last one, large-boilered 6004, was the only one owned by BR, and was scrapped in 1949.

An LNWR 4-6-0 that was given more consideration was Bowen-Cooke's 'Prince of Wales' developed from George Whales' 'Experiment' 4-6-0 of 1905. No 245 had been constructed between 1911 and 1922. They were two inside-cylinder engines, with 6ft 3in coupled wheels, and Joy's valve gear, had 175lbs psi boiler pressure, were superheated and had a tractive effort of 21,760lbs. As mentioned in a previous chapter, No 388 of the class was tested against the Midland's 998 and 1008 in December 1923, and although it matched the performance of the Midland engines on the road between Leeds and Carlisle, its coal consumption was heavier.

The passenger engines of the Scottish railways seem to have been given scant consideration. Probably the best were the Pickersgill

George Hughes' 'Dreadnought' 4-6-0 built in 1921 to the 1919 rebuilding design of the 1908 four-cylinder L&Y 4-6-0, c1925. (Real Photographs/MLS Collection)

Charles Bowen-Cooke's 4-cylinder 'Claughton' of 1913 design, No 208, converted to oil-burning during the 1921 miners' strike, at Camden. It was renumbered 6024 by the LMS. (MLS Collection)

Two Pickersgill '72' class locomotives, No 14503 (built in 1921) behind, outside Glasgow Central station, c1922. The heaviest and fastest Caledonian expresses were double-headed by a pair of McIntosh or Pickersgill Dunalastair 4-4-0s. *(F. Moore/ MLS Collection)*

class '72' 4-4-0s developed from the McIntosh Dunalastairs, which in turn were really developments of Drummond's original 4-4-0s of 1884. The prestige passenger engines of the Caledonian Railway were the 'Cardean' 4-6-0 class, but they were few in number and of 1902/6 design, superb in appearance and publicity, theoretically powerful, but unremarkable in performance and subject to frequent overheating problems. Pickersgill also built a number of class '60' 4-6-0s, with 180lbs psi boiler and 6ft 1in coupled wheels, but these were used more on heavy mixed traffic duties with the Dunalastair 4-4-0s retaining most of the CR's express work. The class '72s' had 6ft 6in coupled wheels, 180lbs psi boilers and had a tractive effort of 21,435lbs. The Highland Railway had the Peter Drummond 'Castle' 4-6-0s built between 1900 and 1917 for the heavy gradients of the Highland main line rather than fast running. Christopher Cummings 'Clan' 4-6-0s of 1919 which looked impressive but with 6ft coupled wheels were equally best suitable for heavily graded lines.

The Glasgow & South Western Railway's expresses were still in the hands of Manson 4-6-0s dating from 1910, with just a few Drummond and Whitelegg 4-4-0s and seventeen Manson 4-4-0s rebuilt in 1920-21. An interesting survivor was the first 4-cylinder engine built in Great Britain, a 4-4-0 numbered 394, and named *Lord Glenarthur*, which had been successively rebuilt by Drummond, then Whitelegg in 1922, renumbered 14509 by the LMS.

Henry Fowler had been a member of the Association of

Pickersgill 4-6-0 No 63, used for heavy passenger and mixed traffic over the Caledonian system, c1920. The LMS later built a further twenty of these engines with increased cylinder size in 1925 numbered 14630-14649. (F. Moore/MLS Collection)

Peter Drummond's Highland Railway 'Castle' 4-6-0, No 140 *Taymouth Castle* (later renumbered 14765), at Inverness roundhouse, built for the Perth-Inverness route, c1920. (F. Moore/MLS Collection)

Cumming's impressive-looking 4-6-0, No 55 *Clan Mackinnon*, thought to be at Perth, c1922. (F. Moore/MLS Collection)

Manson's outside cylindered 4-6-0 of 1910, No 14674, c1928. (Photomatic/MLS Collection)

Manson's 4-4-0 No 394 *Lord Glenarthur* of 1897, the first four-cylindered British locomotive, substantially rebuilt by Peter Drummond and then later in 1922 by Whitlegg. It remained a solitary example. (Photomatic/MLS Collection)

Railway Locomotive Engineers (ARLE) in the First World War, a design committee chaired by Richard Maunsell, and in 1918 the group, which had also included Churchward, produced drawings of a number of potential post-war standard designs as the nationalisation of the railways was then being actively considered. Derby produced two passenger designs in 1918, a 4-4-0 and a 4-6-0 (see Appendix, page 342). The 4-4-0 looked similar in outline to the Midland 483 class but it is unclear whether Maunsell had an influence to adopt long lap/travel valves as on his SECR 'D' and 'E' rebuilds. The 4-6-0 appears to be a powerful two inside cylinder locomotive, an extension of the Derby 4-4-0 design. These options therefore must also have been in Henry Fowler's mind as he and Hughes considered the future.

Two main events shaped the future motive power policy of the LMS. The first was the infiltration of former Midland Railway officers into key roles in the LMS general and motive power management posts. Sir Guy Granet, ex-Midland General Manager, was the first LMS President and Chairman of the Board. Cecil Paget's 1909 Control systems and rigid train loading policies swept LNWR systems away. The appointment of James Anderson in the Traffic Department overseeing train planning and motive power on the road, with the former Midland policy of frequent light and fast express services, was preferred to the less frequent heavy West Coast trains hauled with mighty effort not just by the Claughtons and Prince of Wales 4-6-0s, but by the robust 'George V' 4-4-0s – but all at the expense of high coal consumption and wear and tear through the constant 'thrashing' to keep time with 400+ ton loads. 'George V' 4-4-0 No 1595 was credited with 1,300 dbhp with a 410 ton train. There was much managerial infighting between officers of the former LNWR and Midland railways, with the Midland men coming out on top. The commercial department continued to be subservient to the operators. The Midland's crimson lake livery replaced the LNWR lined black (at least until BR days). The second was the performance of the Midland 4-4-0s, particularly that of Compound 1008 in the tests that took place in 1923, with its superior coal consumption, an important element in the thinking of the cost-conscious LMS management. This was further consolidated by more locomotive trials in 1924, with newly built Compounds 1065 and 1066 taking the honours in general economy against a Claughton, Prince of Wales and Caledonian class 60 4-6-0.

George Hughes had retired aged 60 at the end of 1924, possibly disappointed in, and lacking support from, the new management. His replacement was Henry Fowler who had conceived the idea of a 3-cylinder compound 4-6-0 and the Derby drawing office under his direction had produced a drawing of such a machine in 1924 (see Appendix, page 342), but the die had already been cast and the Midland small engine policy held priority. However, faced with the increasing loads on the main London-Scotland West Coast route demanding more than double-headed Compounds, Fowler drew up designs for a 4-cylinder compound pacific in 1926 (see Appendix, page 343). The LMS operators led by James Anderson were still wedded to the small engine policy – or at least, as small as would do the job – and Fowler's 4-6-0 'simple' Royal Scots were hurriedly built under contract after the abortive effort to obtain fifty Castles built to the GWR design, which had demonstrated to LMS management that a 4-6-0 could do all that was necessary. In the interim, the Fowler version of the Midland Compound was multiplied to 195 locomotives and, as late as 1928, the LMS settled for 138 2P 4-4-0s for lighter and semi-fast services, straight developments of the Midland superheated '483' class of 1912, whose origins dated back to the 1870s.

Chapter 6
THE LMS 4P COMPOUND

Design & construction

Although the LMS embarked on trials with express engines of the constituted companies between December 1923 and 1925, it would seem that the former Midland officers had already decided to build a version of the Deeley Compound as superheated by Fowler. The continuing programme of superheating the Midland Compounds was in progress and as early as June 1923 an order for twenty new locomotives was made, although with George Hughes in charge, a completion of the order of the L&Y 4-6-0s was being completed too. At this time, Sir Henry Fowler (he was knighted after the war in 1918) was acting as Hughes' assistant, although he must have retained considerable autonomy at Derby and encouraged by James Anderson, now Superintendent of Motive Power for the whole company, anticipated the likely results of the trials.

These twenty Fowler Compounds were delivered in 1924 and had a number of relatively small and detailed changes from the Midland superheated Compound design, the most significant of which was a reduction in coupled wheel diameter from 7ft to 6ft 9in.

The principal dimensions of the LMS Compounds were:

Cylinder diameter	
– high pressure	19¾in x 26in (adjusted to 19in)
– low pressure	21¾in x 26in (adjusted to 21in)
Stephenson's Link Motion	
Coupled wheel diameter	6ft 9in
Bogie wheel diameter	3ft 6½in
Boiler pressure	200lbs psi
Heating surface	1,607.7sq ft (of which superheater 290.7sq ft)
Grate area	28.4sq ft
Axleload	19¾ tons
Weight	
– Engine	61 tons 14 cwt
– Tender	42 tons 14 cwt
– Total	104 tons 8 cwt
Water capacity	3,500 gallons
Coal capacity	5½ tons
Tractive effort	22,649lbs

Other detailed differences were:

Chimneys: tall Fowler chimney on 1045-1064 replaced by tall Stanier chimney but the boilers were changed at overhaul to different locomotives. The rest were built with short Fowler chimneys, most replaced by short Stanier chimneys.

Safety valves: 1045-1064 had Ramsbottom valves, later replaced by Ross Pop valves and all the rest built with Ross safety valves.

Drive: 1045-1064 built for right-hand drive similar to the Midland engines. The remainder had left-hand drive.

Tenders: Midland Deeley six-wheel tenders for 1045-1064. New Fowler six-wheel tenders without coal rails were built for the remainder but many tender changes took place during the life of the class.

After the success of the Midland Compound was demonstrated in December 1923 and in line with the LMS policy of small engines and light frequent trains espoused by James Anderson, a further 170 Compounds were built between 1924 and 1927, with then a gap until 1932 when a further five were built with a clear intention to build more (viz. the decision in 1928 to renumber the ten 990 class to provide room for more Compound numbers). The second batch,

The first Fowler LMS Derby built Compound, 1045, with tall chimney and 6ft 9in coupled wheels, in Works grey livery, 1924. (F. Moore/MLS Collection)

1065-1084, were also built at Derby in 1924, and in order to accelerate the delivery of the large number ordered, the 1925 constructed engines were split over four Works:

Derby:	1085-1114
Horwich:	1115-1134 (last four completed in 1926)
North British Co:	1135-1159
Vulcan Foundry:	1160-1184

Vulcan Foundry were contracted to build a further series and 1185-1199, and 900-934 were built and delivered in 1927. A gap of five years then elapsed before a further order was made and 935-939 were built at Derby in 1932, before the new CME, William Stanier, immediately put a stop to further construction. These later engines had enlarged frames and many Compounds were strengthened in this way before withdrawal. It is remarkable to think that at the time the LMS was building Compound 4-4-0s for the main LMS express passenger work, the LNER was building Pacifics, the Great Western Castle and King 4-6-0s and the Southern King Arthurs and Lord Nelsons. Only with apparent reluctance after the demonstration of the GWR Castle's superiority on the West Coast main line in 1925, did the LMS authorities have a rethink, and a hurried design produced the Fowler Royal Scot for the Euston-Crewe-Scotland main services. The Midland main line services remained in the hands of the Compounds.

Few subsequent changes were made. A number were oil fired during the 1926 General Strike, including 1050, 1059, 1066, 1069, 1098, 1104, 1117, 1121, 1124,

The open smokebox door and superheater units of an LMS Compound, c1925. (E.M. Johnson Collection)

No 1112 built at Derby in 1925 in Works grey livery. It has the smaller Fowler chimney. (Railway Photographs/MLS Collection)

1137, 1139 and 1149. No 934 was experimentally fitted with a Westinghouse compressor and air reservoir in 1930-31. Three Deeley tenders were rebuilt with extended coal rails to increase coal capacity to nine tons, one of which being used for 1054's non-stop run from Euston-Edinburgh in 1928 as a publicity stunt to challenge the LNER commencement of non-stop running on the East Coast. The following engines had these tenders at different times: 1053, 1054, 1057, 1059 and 1061. One Fowler 3,500 gallon tender was rebuilt in 1933 with a high-side curved top (prototype for the standard Stanier LMS tender?). This was initially attached to 936 and later, in 1954, to fellow Monument Lane based 933. A few in addition to the first twenty received boilers with tall Stanier chimneys. No 1074 kept it until withdrawal. Others that had such chimneys for periods included 936, 1111 and 1167.

All were taken into BR ownership in 1948 and were renumbered 41045-41199 and 40900-40939. All except the 1932 five were delivered in the standard LMS livery of Crimson Lake with black and yellow lining with number on the tender. The number was moved to the cab later, from around 1928, as confusion arose as tenders were often exchanged during repairs. Nos 935-939 had their numbers on the cab from the beginning. During the Second World War, nearly all received plain black. No 934 was the last to retain the Crimson Lake livery – it was still red at nationalisation. The Compounds were awarded the BR livery of mixed traffic lined black (an irony that after all the intercompany infighting of 1923-4, especially between former LNWR and Midland officers, that these 'Midland' inspired engines finished up in the former LNWR colours!).

The first LMS Compound to be withdrawn was Dumfries' 41171 in December 1952. Nos 41109, 41182, 40911, 40918 and 40922 were also withdrawn at the end of the year. Withdrawals then took place steadily throughout the 1950s, the annual rate being:

1953: 13
1954: 23
1955: 37
1956: 27
1957: 34
1958: 36
1959: 13
1960: 40907, 41063, 41157, 41162
1961: 40936 (January), 41168 (July)

Derby built 1111 at Manchester Longsight shed, in Crimson Lake livery with large numerals on the tender, 1925. (W. Potter/MLS Collection)

North British built Compound 1137 delivered new in 1925. (MLS Collection)

North British built Compound 1152 at Camden, c1926.
(MLS Collection)

Vulcan Foundry built Compound 923, constructed in 1927, at Perth, 29 May 1930.
(H.C. Casserley/MLS Collection)

No 936, one of the five built in 1932, and equipped with the prototype high-sided tender, at Derby, c1935. It also has a tall chimney as seen on the first twenty LMS Compounds. It has an exhaust steam injector and the extension of the frame in front of the smokebox is clear. (MLS Collection)

No 1061, of the first 1924 batch of twenty Compounds, equipped with one of the three Deeley tenders with extended coal rails, c1932. (F. Moore/ MLS Collection)

No 1072 with Midland 2-4-0 No 216, in the Derby Paint Shop, May 1933. (Photomatic/MLS Collection)

No 1086 at Leeds Holbeck, 26 August 1934. Cleaning standards have clearly dropped by the 1930s, the number is now repainted on the cabside with LMS on the tender. (W.L. Good/MLS Collection)

No 1157 at Leamington Spa with a stopping train for Coventry and Birmingham, attached to a Fowler tender with coal rails, and with outside steam pipes, 1937.
(G.A. Coltas/MLS Collection)

M935, one of the last five 1932 built Compounds at Derby to a rougher finish (the numerous rivets very obvious) and given the prefix M before BR renumbering, seen at Derby shed, 1948.
(MLS Collection)

M1162 in early BR livery with the M prefix, at Chester shed, 26 March 1948. I had a number of runs behind this engine in 1957 when it was shedded at Rugby and worked semi-fast services during the summer months to Euston.
(J.D. Darby/MLS Collection)

No 41194 in early BR livery with BRITISH RAILWAYS inscribed on the tender and the cabside showing recent signs of repaint with its BR number, at Accrington, 25 July 1948. This engine, like many others in later years, received a strengthened frame extension and exhaust steam injector.
(H.D. Bowtell/MLS Collection)

No 41085 of Bolton at its home depot in typical early/mid 1950s condition, c1952.
(J. Davenport/MLS Collection)

No 40920 of Ayr at Glasgow St Enoch in plain black and tender without any marking at all, 3 May 1952. This locomotive has the frame extension, but not exhaust steam injector.
(J. Robertson/MLS Collection)

Kettering's 41192 at Derby in the condition and livery most will remember the Compounds bore in the 1950s, 21 April 1954. (MLS Collection)

No 40903 in store at Carstairs with a Dunalastair, 7 September 1952. No 40903 was a Carstairs based engine and remained there until eventually withdrawn in 1955. (N. Harrop/MLS Collection)

Operations

The performance of Midland Compound 1008 in December 1923 and January 1924 confirmed the LMS's new management in their preference for the Midland Compounds as a basis for new construction. No 1008 had a particularly strong reputation and full logs of its performance are on pages 201 and 202, where its log is shown alongside that of the competing 998 and LNWR 388. Whilst the 'Prince of Wales' matched 1008's times or nearly so, the advantage of the Compound lay in the coal consumption figures. No 1008 consumed 3.64-4.02lbs per dbhp with 300 tons and 3.7-4.02lbs with 350 tons, recording 1,000+ horsepower at 50-65mph. The LNWR engine kept time but with coal consumption varying from 4.31-5.12lbs per dbhp.

The Euston-Birmingham fast trains were relatively lightly loaded (normally six coaches) and, in 1924, Anderson arranged for a Midland class 2 and Compound 1033 to be tested on this route. The 2P struggled to keep time but 1033 performed satisfactorily and after 1045-1050 had been allocated to Midland line sheds, 1051-1056 went to Camden for the Birmingham services.

Further tests took place in the autumn of 1924 on the Settle & Carlisle line, comparing the new 6ft 9in Compounds against the Caledonian Pickersgill 4-4-0 and an LNWR 'Claughton', theoretically more powerful. Nos 1065 and 1066 represented the LMS Compounds, 1023 the Midland Compounds and the 'Claughton' was a 'run of the mill' engine from Edge Hill depot that could be spared, 2221 *Sir Francis Dent*. After a couple of poor trips with the Dunalastair which proved incapable of timing the 300 ton loads, it was sent back to Scotland whilst the 'Claughton' and Compounds tackled 350 tons. The trial was biased and proved to Fowler and Anderson what they wanted to hear. The coal consumption of the two LMS Compounds was good, consistent at around 4.35-4.45lbs per dbhp per hour, though not as good as 1008's 1923 exploits described above. No 1023 produced the same economy figures but had some poor performances, especially on the banks with the 350 ton loads. No 2221 lost some time on the banks but regained it with fast downhill running, but its coal consumption was heavy at 5.56lbs per dbhp hour.

No 1065 on 20 November 1924 took 361 tons tare, 385 gross, on the northbound service and completed the 14 mile climb from Settle Junction to Blea Moor in 25 minutes 47 seconds, with a minimum of 25mph on the climb. No 1066 on 9 December with 305 tons tare took just 23 minutes and 5 seconds, not falling below 30mph. With 347 tons tare the following day, 1066 took 23 minutes 58 seconds holding 28½mph at the summit. The maximum horsepower recorded was 906 with 1066 on the 350 ton load (less than 1008 had recorded in the previous year).

As the Compounds were delivered, they were allocated all over the LMS territory. The initial distribution by Motive Power Division, was:

Midland:	1000 -1044 (Deeley engines), 1045-1050, 1057-1064, 1070-1075, 1085-1109.
LNW:	1051-1056, 1076-1084, 1110-1134, 1150-1184.
Scottish:	1065-1069, 1135-1149.

In February 1925, Derby tested two LMS Compounds with the dynamometer car on the Settle & Carlisle which had experimental alterations to address the complaint that sometimes on the heavier working the cylinders 'beat the boiler'. No 1060 was fitted with a reduced blast pipe cap diameter and petticoat, while 1065 had its cylinders lined up to 19 and 21 inches. Whilst 1060 steamed freely, 1065 was able to do the same work with less regulator and cut-off. Despite the smaller cylinders, 1065 worked less hard, gained time on the gradients with 350 tons and achieved the work with coal consumption figures of 4.15-4.22lbs per dbhp per hour compared with 1060's 4.65-4.77. Analysis showed that the LMS Compounds were at their best with uphill heavy work at 30-40mph. At higher speeds, the front end throttled the steam and they seemed better if the driver eased the throttle back. This supports the working in Midland Railway days when observations indicated that the Compounds went hard uphill but were not usually pushed down hill above 70-75mph, unless trying to regain lost time.

With the authorities' intention to spread the use of the Compounds to all parts of the LMS system, further trials were carried out in May 1925, but this time between Preston and Carlisle. The tests were on special services of empty stock rather than scheduled passenger trains, unlike the previous tests. Four locomotives were tested – a Hughes rebuilt 4-6-0 10460, LNW 'Claughton' 30 *Thalaba* (in good condition this time), and a Prince of Wales 4-6-0, as well as LMS Compound 1065. A Caledonian class 60 4-6-0 was tested similarly in 1926. The Compound had run 28,000 miles since its last overhaul, and 10,000 miles since its cylinders had been lined up for the previous Settle & Carlisle trials. The 'Prince of Wales' and the Compound were power-classified '4' by the LMS authorities and the other two 4-6-0s as '5'. The '4s' took two trains of 300 and 350 tons, the '5s' trains of 350 and 400 tons. The northbound trains ran to the schedules of the best advertised services – 103 minutes for the 90 miles but with substantially more coaches than that now laid down for unassisted engines over Grayrigg and Shap – 260 tons. The best runs for which we still have details were:

Preston-Carlisle Test runs, 1925

Miles	Location	1065 LMS Compound 347 tons			10460 Hughes 4-6-0 397 tons			30 *Thalaba* Claughton 4-6-0 394 tons			Gradients
		Times	Speeds		Times	Speeds		Times	Speeds		
0	Preston	00.00			00.00			00.00			
1.3	Oxheys Box	04.24	pws	1½ L	04.09	pws	1 L	04.05	pws	1L	
21	Lancaster	24.35		1½ L	25.48		2¾ L	25.53		3 L	
27.3	Carnforth	30.49		1¾ L	32.17		3¼ L	32.10		3¼ L	
40.1	Oxenholme	43.40		¾ L	46.00		3 L	46.28		3½ L	1/111 R
47.2	Grayrigg	53.27	43 ave	1½ E	55.41	44 ave	¾ L	57.49	37 ave	2¾ L	1/106 R
53.2	Tebay	59.48	57 ave	1¼ E	61.45	60 ave	¾ L	64.42	52 ave	3¾ L	L
58.7	Shap Summit	68.14	25½	2¾ E	70.27	24	½ E	73.39	28 ½	2¾ L	1/75 R
	Southwaite	-	70		-	74		-	79		1/172 F
90.1	Carlisle	100.00		3 E	101.04		2 E	102.16		¾ E	
		(98 net)			(98½ net)			(99½ net)			

All performed well, but the efforts of the class '5s' were at the expense of higher coal consumption. The average figures calculated at the end of the trials were:

	Coal consumption lb per mile	Average per dbhp hour
'Claughton' 30 *Thalaba*	42.4, 52.7 (latter with 400 tons)	4.78 (with 350 tons)
1065	43.4	4.25
'Prince of Wales' 90 *Kestrel*	48.3	5.05
Hughes 10460	51.3, 58.1 (latter with 400 tons)	5.07
CR 60 (1926)	51.55	4.84

One further trial took place in 1926 when the LMS obtained the loan of GW 5000 *Launceston Castle* which demonstrated on the West Coast route what a 4-6-0 with good front end design could do with loads that would require pilot assistance on the LMS timing policy. In a rather unequal exchange, Compound 1047 went temporarily to the Great Western and was used mainly between Paddington and Bristol on relatively fast but lightly loaded trains. It was not trusted with the loads then being conveyed by the 'Castles' to Devon and Cornwall. One recorded run on the fast 1.15pm Paddington-Bristol was timed by C.J. Allen. With seven coaches, 210 tons gross, 1047 accelerated to 68mph before a p-way slowing at Slough, then sustained 70-72 from Reading to Didcot falling away to 60 at Uffington before a two-minute dead stand for signals at Shrivenham. In an attempt to recover time 1047 was pushed to 76½mph down Dauntsey Bank and 80 at Box, and arrival at Bath on the 105 minute 60mph schedule was four minutes late but only 100 minutes net.

During the years 1925-1928, the *Royal Scot* and other Anglo-Scottish expresses were hauled by pairs of Compounds both north and south of the border. By the end of 1927, the first 'Royal Scot' 4-6-0s were appearing and by the end of 1928 they had taken over from the Compounds on the main West Coast services north of Crewe. In 1926, the General Strike and prolonged miners' strike caused a number of the LMS as well as Midland Compounds to be converted to oil-burning, one of which was the Scottish Division based 1069 which features in runs tabled below on the G&SWR route from Carlisle to Glasgow St Enoch. The logs featured in Cecil J. Allen's columns of the January 1928 edition of the *Railway Magazine*, though it is not clear whether 1069 was still oil-fired at the time.

Carlisle-Glasgow St Enoch, c 1926/7

Miles	Location	1069 188/200 tons Times	Speeds		1068 214/225 tons Times	Speeds		1069 223/235 tons Times	Speeds		Gradients
0	Carlisle	00.00			00.00			00.00			
4.1	Rockcliffe	06.20	62		06.35	60		06.25	60		
8.7	Gretna Jcn	10.53		1¼ E	11.05		1 E	10.55		1 E	
11.7	Rigg	14.09			14.35			14.10			
14.6	Eastriggs	17.16	58		17.35	60		17.00	62		
17.6	Annan	20.13	62	1¾ E	21.00		T	20.15		¾ E	
21	Cummertrees	23.30			05.35			05.30			
24.6	Ruthwell	27.13	56		09.35			09.15	57		1/200 R
29.2	Racks	31.50	62		14.35			13.55	62		1/150 F
33.1	Dumfries	40.43		¼ E	19.35		½ E	18.50		1¼ E	
0		00.00			00.00			00.00			
3.4	Holywood	05.30			05.55			05.50			
7.6	Auldgirth	10.01			10.45			10.45			1/200 R
11.5	Closeburn	14.24	53		15.30	49		15.40	47½		1/200 R
14.2	Thornhill	17.37	50	½ E	19.00	46	1 L	19.15	45	1¼ L	1/150 R
17.5	Carronbridge	21.44	48		23.50	41		24.00	42		1/150 R

The LMS 4P Compound

Carlisle-Glasgow St Enoch, c 1926/7

Miles	Location	1069 188/200 tons Times	Speeds		1068 214/225 tons Times	Speeds		1069 223/235 tons Times	Speeds		Gradients
26.1	Sanquhar	30.52		2 E	34.10		1¼ L	34.20		1¼ L	1/200 F, 1/180 R
29.5	Kirkconnel	34.15	60		38.15			38.05			1/200 R
36.9	New Cumnock	42.16		4¾ E	46.10		¾ E	46.40		¼ E	L
42.3	Old Cumnock	48.24			52.35			52.15	60		1/145 F
44.3	Auchinleck	50.37			55.51			54.35			1/180 F
48.7	Mauchline	55.10		4¾ E	60.40		¼ E	59.15		1¾ E	1/150 F
56.3	Hurlford	62.31			69.45			67.28			1/100 F
58.1	Kilmarnock	64.50		7¼ E	72.25		½ E	70.00		3 E	
0		00.00			00.00			00.00			
2.3	Kilmaurs	04.27			04.45			04.55			1/180 R
5.5	Stewarton	10.03	34		09.35	40		10.20	35½		1/87 R, 1/152 R
7.7	Dunlop	14.29	30		13.15	36		14.50	29		1/75 R
10	Lugton	17.19		¼ L	15.55		T	17.35		1½ L	
11.6	Caldwell	19.16			17.50			19.25			L/1/100 R
14.7	Neilston	22.47			21.15			22.50			1/69 F
16.8	Barrhead	25.06		T	23.35		½ E	24.55		1 L	1/100 F
18.5	Nitshill	-	sigs		25.30			26.50			
22.5	Strathbungo	-	sigs		29.40			31.20			
24.4	Glasgow St E	43.09		2¼ L	33.20		1¾ E	35.10		¼ L	

Glasgow St Enoch-Carlisle, c1926/7

Miles	Location	915 220/230 tons Times	Speeds		1149 + 14476* 260/275 tons Times	Speeds		1065 +14317** 300/320 tons Times	Speeds		Gradients
0	Glasgow St E	00.00			00.00			00.00			
1.9	Strathbungo	04.35		½ L	03.51		T	03.50		T	
5.9	Nitshill	09.45	51		08.51	55		08.51	51		
7.6	Barrhead	11.40		¾ L	10.40		¼ E	10.52		T	1/100 R
9.7	Neilston	15.00	34		13.47	41		14.18	37		1/69 R
14.4	Lugton	22.20		¾ E	20.48		2¼ E	21.06		2 E	

		Glasgow St Enoch-Carlisle, c1926/7							
		915			**1149 + 14476***		**1065 +14317****		
		220/230 tons			**260/275 tons**		**300/320 tons**		
Miles	Location	Times	Speeds		Times	Speeds	Times	Speeds	Gradients
18.9	Stewarton	27.40	60		26.39		26.38		1/75 R
22.1	Kilmaurs	30.45	63		29.47	60	29.52	58	1/87 F
24.4	Kilmarnock	33.50		¼ E	32.42	1¼ E	33.03	1 E	
0		00.00			00.00		00.00		
1.8	Hurlford	03.50			03.38		04.00		
9.4	Mauchline	15.10	41	¼ L	15.37	38 ½ L	15.55	39 1 L	1/100 R, L
13.8	Auchinleck	20.15	52		20.47	50	21.37	46	1/180 R
15.8	Old Cumnock	23.05			23.08		24.20		
21.2	New Cumnock	30.20	44 ½	2¾ E	29.43	50 3¼ E	31.22	46 ½ 1½ E	1/175 R, L
28.6	Kirkconnel	38.25	pws		37.58		39.01		1/200 F
32	Sanquhar	42.00		2 E	41.15	58 2¾ E	42.19	62 1¾ E	L
40.6	Carronbridge	52.05	easy		49.58	60	51.12	easy	1/180 F
43.9	Thornhill	55.55		1 E	54.18	3¾ E	55.10	2¾ E	
46.6	Closeburn	58.45	60						1/150 F
50.5	Auldgirth	62.50	58 easy						1/200 F
58.1	Dumfries	71.35		½ E					

* 14476 was a Pickersgill 4-4-0
** 14317 was a McIntosh Dunalastair 4-4-0

No 915 then ran easily, stopping at Annan, and arrived at Carlisle just before time, without exceeding 60mph. A pair of Compounds, 914 and 1065, with a gross load of 380 tons ran non-stop from Dumfries to Carlisle, 33.1 miles, in 39 minutes 50 seconds (39 net) with a top speed of 68mph on the level through Annan.

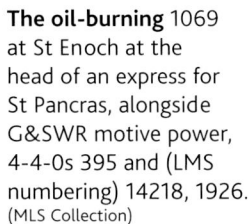

The oil-burning 1069 at St Enoch at the head of an express for St Pancras, alongside G&SWR motive power, 4-4-0s 395 and (LMS numbering) 14218, 1926. (MLS Collection)

The Anglo-Scottish trains at this time were usually worked by pairs of Compounds both north and south of Carlisle, and tabled below is a run on the northbound *Royal Scot* logged around 1927.

Carlisle-Symington, c1927
The Royal Scot
1133 + 1139
415/435 tons

Miles	Location	Times	Speeds		Gradients
0	Carlisle	00.00			
4.1	Rockcliffe	06.09	64		L
8.6	Gretna Junction	10.18		¾ E	
13	Kirkpatrick	15.19	53		1/200 R
16.7	Kirtlebridge	19.30		½ E	L
20.1	Ecclefechan	23.00	58		1/200 R
22.7	Castlemilk	25.51			1/200 F
25.8	Lockerbie	28.46	65	1¼ E	1/528 F
28.7	Nethercleugh	31.20			
34.5	Wamphray	36.48	63		1/326 R
39.7	Beattock	42.06	59	2 E	1/202 R
45.4	Greskine Box	51.36	30		1/79 R, 1/74 R
47.8	Harthorpe Box	56.26			1/75 R
49.7	Summit	60.31	28	3½ E	1/74 R
52.6	Elvanfoot	64.00			1/99 F
57.8	Abington	69.00	62½		1/392 F
63.2	Lamington	73.59	65		1/200 F
66.9	Symington	78.32		2½ E	(detach Edinburgh portion)

A typical Compound performance, hard uphill but taken very easily downhill. Southbound 903 and 905 were timed from Symington to Carlisle with a full 445 ton train (470 gross) and after clearing Beattock summit at 38mph, ran up to 74 on the long descent. The pair worked through to Carnforth and after passing Carlisle on time, gained three minutes on the schedule in the climb to Shap, passed at 37½mph, and were then allowed to reach 77½mph through Tebay before easing to under 50mph before the descent from Grayrigg was commenced, touching 75 at Oxenholme. Carnforth was reached six minutes early.

In 1928, the LNER inaugurated the non-stop *Flying Scotsman* between London and Edinburgh and the day before the LMS stole a march by running the *Royal Scot* in two portions, the Glasgow train with 'Royal Scot' 6113, but the Edinburgh six-coach portion with Compound 1054 specially equipped with the extended tender to take nine tons of coal. The load was light and the timings required no special effort other than judging the run within the coal and water capacity.

Whilst the Compounds seem to have been well received in Scotland, the engines allocated to the Western former LNWR section had a very mixed response. The Midland Paget/Anderson policy

No 1054, the locomotive that hauled the Euston-Edinburgh portion of the *Royal Scot* non-stop in 1928, the day before the LNER *Flying Scotsman* commenced its well-publicised non-stop running to Edinburgh. It was provided with a tender with enhanced coal capacity for the run and is seen here on a Scotch express near Carpenders Park, 1928. (Railway Photographs/MLS Collection)

No 1054 on an up Scotch express near Bushey, c1929. It and the other Compounds from the 1045-1064 series are still seen with the tall chimneys in this set of photographs from the 1920s. (Real Photographs/MLS Collection)

of rigidly restricted loads meant much piloting of the heavy West Coast expresses, where previously the 'Claughtons' and the 'George V' 4-4-0s had managed perfectly well with 400 tons or more, the single engine load without assistance for both class 5s (Hughes and Claughton 4-6-0s) and 4s (Compounds and 'Prince of Wales') being just 260 tons. Loads on the Euston-Birmingham route were however within the loading set for a single engine and the first group of Compounds allocated to the Western Division, 1051-1056, were used on these services. The *Railway Magazine* of May 1930 tabled a number of runs with these engines and I've selected the four below:

The West Coast main line south of Rugby has four long gradients of six to eight miles at around 1 in 330 to Bushey, Tring, Roade and Welton and the similar descents. The Compounds climbed these at around 55-60mph but were not tempted to do more than 70mph on the downhill stretches, though all were within scheduled time, with 1053 on the fourth run having one coach more than that stipulated without assistance. Not tabulated in the magazine article was a further run with 1164 which had two extra coaches, 301/320 tons, and this could not quite keep the schedule losing just over a minute without any checks. In the early 1930s, when the Birmingham services were the

only regular express workings they retained on a regular basis, the Compounds were diagrammed for the tightly timed two-hour 4.50pm from Birmingham which included stops at Coventry and Willesden Junction, and was allowed 87 minutes for the 88.6 miles from Coventry to Willesden. No 1134, with 290 tons gross, covered the ground in 84 minutes 36 seconds (82 net) with 73mph before Rugby, 77½ at Weedon, 80 at Castlethorpe, 63 at Tring and 77½ at King's Langley. No 1111 on the 4.35pm down with 280 tons gross reached its first stop at Blisworth (62.2 miles) in 61 minutes, 4 seconds (59½ net), with 60mph minimum at Tring, 80 at Sears Crossing, 77½

Euston-Coventry, c1928/9

Miles	Location	1171 231/245 tons			1056 253/265 tons			1054 256/270 tons			1053 286/300 tons		
		Times	Speeds		Times	Speeds		Times	Speeds		Times	Speeds	
0	Euston	00.00			00.00			00.00			00.00		
1	Camden	03.06			03.24			03.01			03.49		
5.5	Willesden Jcn	09.15		¼ L	09.29		½ L	09.09		¼ L	11.15	pws	2¼ L
17.5	Watford	21.17	60	¾ E	21.01	63	1 E	21.50	58	¼ E	23.20	60	1¼ L
21	King's Langley	24.37	63		24.15	65		25.26	58		26.47	61	
28	Berkhamsted	31.34	61		30.56	63		33.02	pws		33.55	59	
31.6	Tring	35.21	56	2¾ E	34.41	60/58	3¼ E	38.13	41	¼ L	37.55	54 T	
40.3	Leighton B'zz	42.43			42.05			46.02			45.33		
46.6	Bletchley	48.18	67½	2¾ E	47.39	69	3¼ E	51.42	67	¾ L	51.09	67½	¼ L
52.4	Wolverton	53.24	68		52.39	70		56.41	70		56.10	70	
59.9	Roade	60.40	58	2¼ E	60.08	59	2¾ E	63.37	64	½ L	63.13	62	¼ L
62.8	Blisworth	63.34	65	2½ E	63.10	60	2¾ E	66.19	72	¼ L	66.01	70 T	
69.8	Weedon	70.01	64½		69.33	66		72.18	70		72.01	70	
75.3	Welton	76.29	pws		74.58	61		77.21	59		77.06	58	
82.5	Rugby	84.18		¾ E	82.18		2¾ E	84.35		½ E	86.05	sigs	1 L
89	Brandon	90.57	60		88.53	62		90.42	64		93.17	pws	
94	Coventry	96.21		¾ E	94.37	sigs	2¼ E	95.42		1¼ E	98.57		2 L

at Wolverton, and 64½ at Roade summit. Speeds above 70mph were becoming necessary to keep time on the accelerated services in the 1930s.

By the delivery of the final five Compounds in 1932, the 'Royal Scots' had replaced the Compounds on the main West Coast expresses apart from the Birmingham road and at the beginning of 1933, the allocation of the class members was:

Shed	Numbers
Camden:	1105-1110, 1112, 1134 (8)
Rugby:	1122, 1152-1156 (6)
Bushbury:	1113, 1161-1169 (10)
Monument Lane:	1128, 1129, 1160, 1170, 1172-1174 (7)
Crewe North:	1115, 1118-1121, 1125, 1133 (7)
Chester:	1111, 1114, 1116, 1150, 1151, 1158 (6)
Edge Hill:	1157, 1159
Longsight:	1130-1132 (3)
Carnforth:	1117, 1123, 1124, 1126, 1127, 1171 (6)
Carlisle Kingmoor:	1065-1069, 1135-1149, 1175-1178 (24)
Glasgow Polmadie:	900-909, 1083, 1084 (12)
Dalry Road:	919, 920
Carstairs:	917
Perth:	921-924, 939 (5)
Aberdeen:	918, 938, 1183, 1184 (4)
Glasgow Corkerhill:	911-914, 1080-1082, 1179, 1180 (9)
Hurlford:	915, 916
Ayr:	910, 1181, 1182 (3)
Bank Hall:	937, 1191-1193 (4)
Southport:	1189, 1190
Blackpool:	1194-1197 (4)
Wakefield:	1198, 1199
Low Moor:	1185-1188 (4)
Kentish Town:	1051-1055, 1074 (6)
Leicester:	1099-1104 (6)
Nottingham:	925-928, 1092-1097 (10)
Derby:	933-936, 1045-1050, 1057-1061, 1098 (16)
Sheffield Millhouses:	1056, 1076-1079 (5)
York:	1091
Trafford Park:	1062-1064, 1089, 1090 (5)
Leeds Holbeck:	1070-1072, 1085-1088 (7)
Carlisle Durran Hill:	929-932, 1073, 1075 (6)

The fleet of forty-five former Midland Compounds (1000-1044) were all stationed at Midland Division sheds.

No 1053, one of the Compounds initially allocated to the Euston – Birmingham services, at Hampton-in-Arden with a Wolverhampton-Euston express, April 1927. (W.L. Good/ MLS Collection)

The LMS 4P Compound • 231

No 1052 on an up Birmingham express on Bushey troughs, c1928.
(F. Moore/MLS Collection)

Kingmoor's two Compounds, 1176 and 1135, double-head the royal train on Beattock bank, July 1932.
(G.W. Shott/MLS Collection)

No 1057, one of the three Compounds fitted with a tender with extended coal rails and enlarged coal capacity, by now transferred back from the West Coast to Kentish Town, near Cromford with a St Pancras-Manchester express, c1930. (MLS Collection)

No 1050 emerging from Headstone Tunnel in the slight dip before resuming the ascent to Monsal Dale and Peak Forest with a St Pancras-Manchester express, c1930. (E.M. Johnson Collection)

The LMS 4P Compound • 233

Derby's 1098 drifting down Lickey bank with a Newcastle-Bristol express, 1926. The route has now been strengthened south of Birmingham to permit larger locomotives including the Compounds. (W.L. Good/MLS Collection)

Corkerhill's 914 departing from Annan with a boat train special for Stranraer Harbour, 1928. (MLS Collection)

Chester's 1150 in full cry on Bushey troughs with a heavy boat train for Holyhead, 1928. (MLS Collection)

Liverpool Edge Hill's 1157 piloting 'Patriot' 5514 *Holyhead*, newly rebuilt from a 'Claughton', on a northbound express passing Rugby station, c1932. (G.A. Coltas/ MLS Collection)

The LMS 4P Compound • 235

No 1185 with an up Midland line express at Elstree, August 1931.
(Photomatic/MLS Collection)

No 1189 among the chimneys at Middleton Junction on the L&Y system between Manchester and Rochdale, c1933.
(G.W. Smith/MLS Collection)

Bushbury's 1166 with a Birmingham line express between Rugby and Coventry, c1932.
(G.W. Smith/MLS Collection)

By the mid-1930s, Stanier's 'Black 5s', 'Jubilees' and Pacifics had removed any remaining regular express work from the Compounds' repertoire, although perhaps they lingered longest in Scotland on the Glasgow-Perth-Aberdeen route and on the former G&SWR lines. O.S. Nock rode on the footplate of 1127, transferred from Carnforth to St Rollox in 1933, on its return Glasgow-Aberdeen diagram. With eleven coaches, 332/355 tons, 1127 had a pilot in the form of a Dunalastair from Aberdeen to Forfar, but 1127 worked the heavy train without assistance after that.

Forfar-Glasgow Buchanan Street, c1934
1127 – St Rollox
11 chs 332/355 tons

Miles	Location	Times	Speeds		Gradients
0	Forfar	00.00			
2.9	Kirriemuir Junction	05.01	53	1 L	1/248 F
5.7	Glamis	07.56	60		1/220 F
7.9	Eassie	10.01	64½		L
12	Alyth Junction	13.52	63½	1 L	L

		Forfar-Glasgow Buchanan Street, c1934			
		1127 – St Rollox			
		11 chs 332/355 tons			
Miles	Location	Times	Speeds		Gradients
16.7	Coupar Angus	18.14	65	1¼ L	L
18.9	Burrelton	20.21	62½		1/266 R
21.2	Cargill	22.32	71		1/160 F
25.3	Stanley Junction	26.16	62½	1¼ L	L
28.3	Luncarty	29.01	69/sig stand		1/125 F
<u>32.5</u>	<u>Perth</u>	<u>41.17</u>		<u>6¼ L</u>	
0		00.00			
2	Hilton Junction	05.50/06.30 sigs stand		9 L	
3.9	Forgandenny	10.10	48½		L
9.6	Dunning	16.36	54½		
13.7	Auchterarder	22.43	41/32		1/100 R
15.3	MP 136 (summit)	26.05	28		1/100 R
<u>15.8</u>	<u>Gleneagles</u>	<u>27.10</u>		<u>7½ L</u>	

Thereafter, 1127 just about held schedule, though it was unable to gain any more time and was nine minutes late into Buchanan Street after a p-way slack at Garnkirk. There was a slight net gain between Forfar and Gleneagles and the Gleneagles-Glasgow time would have been the net scheduled time too, but it had to be worked very hard to achieve this. A run from Glasgow to Perth with 1128 with only 195 tons gross kept time easily with a minimum of 41mph from Stirling out to Kinbuck and 82 down the 1 in 100 from Gleneagles. No 1143 on the 7.09am Aberdeen with 260 tons gross kept time to Perth, but the load was made up to 371 tons tare, 395 gross and the Compound struggled with this on Gleneagles Bank, falling to 19½mph and dropping three minutes on the Perth-Gleneagles schedule.

For the rest of the 1930s, the Compounds found themselves mainly on secondary duties such as semi-fast and stopping services between St Pancras and Bedford/Leicester, the North Wales coast, Derby-Manchester-Sheffield, Leeds-Morecambe and Heysham, Glasgow to the Ayrshire Coast and in the Birmingham-Gloucester-Bristol area. They, like the ex-Midland class 2s, also found themselves on frequent piloting work as the LMS increased the number of express services with strictly limited loads, so that 4-4-0s piloting Class 5s, Jubilees or even Royal Scots became a common sight, a situation very rare on other railways except on heavily graded routes such as Newton Abbot-Plymouth. In the late 1930s, Compounds operated in the Manchester-Liverpool area and 1191, based at Bank Hall, was timed on the 5-coach 3.40pm Liverpool Exchange-Manchester Victoria, scheduled 45 minutes for the 36.5 miles. It just kept time despite a p-way slowing to 40mph preventing a run at the 1 in 90/98 climb to Hindley which the Compound surmounted at 51mph and a succession of signal checks in from Salford. The earlier long 1 in 118 climb to Upholland brought the speed down from the mid-60s to 53mph and the Compound's driver allowed the engine to freewheel up to 70mph between Pendleton and Pendlebury.

A couple of snippets covering some of these routes will be the final logs of main line express work apart from specials such as railtours.

Cheltenham-Birmingham, c1936
936 - Derby
292/310 tons

Miles	Location	Times	Speeds		Gradients
0	Cheltenham Lansdown	00.00		T	
3.8	Cleeve	05.08	62		1/295 F
7.3	Ashchurch	08.08	72	T	1/297 F
9.4	Bredon	09.59	68		1/301 R
12.1	Eckington	12.16	73		1/319 F, L
16.7	Wadborough	16.12	66½		1/301 R
18	Abbots Wood Jcn	17.23	66	¾ E	L
26.5	Droitwich Rd Box	24.58	70½		L
31.1	Bromsgrove	29.40/31.08	1 ¼ E/ 1 E	(banker 7236 0-6-0T to rear)	
33.5	Blackwell	38.19	21	¼ L	1/37¾ R
35	Barnt Green	40.36	46		
38.6	Northfield	44.28	66½		1/301 F
42.2	Selly Oak	48.26	46*/54	½ L	
45.5	Birmingham New St	53.55		T	

Shrewsbury-Crewe, May 1929
1132 - Longsight
353/385 tons (Whitsun relief)

Miles	Location	Times	Speed		Gradients
0	Shrewsbury	00.00		T	
4.6	Hadnall	10.55	24½ /54		1/117 R, 1/155 F
7.2	Yorton	14.10	49½		1/167 R
10.7	Wem	17.45	63		1/208 F
14	Prees	21.00	60		L
18.9	Whitchurch	26.05	56		1/290 R, L
	MP 12	-	53		1/136 R
23.8	Wrenbury	30.55	74/69		1/110 F
28.2	Nantwich	34.40	75		1/97 F
32.7	Crewe	41.15	sigs (40 net)	1¼ L	

I couldn't resist putting this last run in with its 75mph past Nantwich where I live, a point now restricted to 60mph round the curve and level crossings.

Interestingly, I've discovered one log from the Rail Performance Society's archive which features a pair of Compounds working a Midland line express as late as 1939 – although one of the pair was a former Midland engine. The run was timed by D.S. Barrie.

Bedford-St Pancras, 4 April 1939
1.55pm Sheffield-St Pancras
1096 + 1022
10 chs, 289/310 tons

Miles	Location	Times	Speeds		Gradients
0	Bedford	00.00			
8	Ampthill	10.17	58	T	1/200 R
	Flitwick	-	64		L
12.5	Harlington	14.43	60		1/200 R
17	Leagrave	19.17	57		1/200 R
19.6	Luton	21.33	73½ / 67	2½ E	1/176 F, 1/200 R
22.5	Chiltern Green	24.03	76		1/176 F
25.2	Harpenden	26.13	69		1/200 R
29.9	St Albans	30.19	72	3¾ E	
	Napsbury	-	sigs 50*		1/176 F
34.6	Radlett	35.01	55		1/176 F
37.4	Elstree	37.4	51		1/200 R
40.5	Mill Hill	41.16	73		1/176 F
42.9	Hendon	43.08	79	2 E	
48.5	Kentish Town	49.08	52	2 E	
49.8	St Pancras	51.36		2½ E	

No 1053, now transferred back to Kentish Town, leaving St Pancras with a down express, 1937. (MLS Collection)

Derby's 1046, now demoted to stopping passenger train working, at New Mills with a Derby-Manchester train, 4 July 1936. (MLS Collection)

Nottingham's 928 at Chapel-en-le-Frith with a Manchester-Nottingham train, 30 June 1934. (MLS Collection)

The LMS 4P Compound • 241

Nottingham's 1093 with a Manchester-Nottingham semi-fast train at Romiley, 5 August 1933.
(MLS Collection)

Crewe's 1121 pulling out of Manchester London Road with a Euston train, 28 September 1934. In the background, an LNER B17 can just be spotted.
(MLS Collection)

Holbeck's 1087 arriving at Carlisle with the down *Thames-Clyde Express* which it has hauled from Leeds via the Settle & Carlisle line, 20 August 1939. (E.R. Morten/MLS Collection)

Nottingham's 1095 passes Darley Dale with a Manchester-Nottingham express, 2 July 1938. (E.R. Morten/MLS Collection)

Kingmoor's 1136 pilots a Corkerhill 'Jubilee' out of Kilmarnock with a Glasgow St Enoch-Carlisle-St Pancras express, c1936.
(MLS Collection)

Monument Lane's 1157 runs into Kenilworth with a Birmingham-Leamington stopping train, 28 August 1936.
(G.A. Coltas/MLS Collection)

Blackpool's 1195 accelerates from South Shore with the 10.10am Blackpool-Manchester, 20 September 1937. (MLS Collection)

Holbeck's 1072 pilots Stanier 'Black 5' 5279 on an up express nearing Chapel-en-le-Frith, 30 May 1939. (MLS Collection)

Holbeck's 1088 at Chinley with a Sheffield-Manchester stopping train, 29 May 1938.
(E.R. Morten/MLS Collection)

1092 pilots unrebuilt 'Royal Scot' 6138 *The London Irish Rifleman* on a West Coast express at Bourne End, 11 July 1939.
(H.C. Casserley/MLS Collection)

The Compounds suffered during the Second World War as they were complex machines and maintenance became protracted. The main line sheds of Camden, Bushbury, Edge Hill and Carlisle Durran Hill had lost their allocation and sheds like Llandudno Junction, Liverpool Brunswick, Dumfries and Stranraer, Kettering and Bedford, Lancaster and Bradford, Saltley and Bourneville, Accrington and Bolton now had several of the class. The allocation immediately after nationalisation in 1949, before any withdrawals, was:

No 1092 handling a heavy express unassisted at Whitmore between Crewe and Stafford, 1939. (E.R. Morten/MLS Collection)

Bushbury's 1113 at Kenilworth with a Birmingham-Leamington local train, 28 August 1936. (G.A. Coltas/MLS Collection)

Rugby:	41090, 41105, 41152, 41165, 41174 (5)
Monument Lane:	41111, 41115, 41116, 41122, 41151, 41166, 41172 (7)
Crewe North:	41112, 41160, 41167, 41173 (4)
Chester:	40900, 40933, 41098, 41106-41108, 41120, 41121, 41153, 41157, 41158, 41162-41164, 41169, 41170 (16)
Llandudno Junction:	40925, 40936, 41086, 41093, 41114, 41118, 41119, 41123, 41124, 41150, 41161 (11)
Liverpool Brunswick:	40926
Longsight:	41113, 41159, 41168 (3)
Lancaster:	40931, 41045, 41053, 41056, 41080, 41081 (6)
Carlisle Kingmoor:	41129, 41139-41143, 41146 (7)
Dumfries:	40902, 40904, 40912, 41109, 41135, 41171, 41175, 41179 (8)
Stranraer:	41092, 41099, 41127 (3)
Glasgow Polmadie:	40916, 41131
Greenock:	41148, 41149, 41182 (3)
Dalry Road:	40911, 41177, 41178 (3)
Carstairs:	40901, 40903, 40907, 41130, 41136, 41145, 41147, 41180 (8)
Perth:	40921-40923, 40938, 40939, 41125 (6)
Aberdeen:	41134, 41176, 41184 (3)
Glasgow Corkerhill:	40905, 40906, 40909, 40914, 40915, 40919 (6)
Hurlford:	41110
Ayr:	40908, 40920, 41132, 41133, 41138, 41155, 41183 (7)
Balornock (St Rollox):	40918, 41126, 41128
Stirling:	40913, 40924
Bank Hall:	40937
Southport:	41085, 41193
Accrington:	41100-41102, 41187, 41188, 41194 (6)
Blackpool:	41185, 41192, 41195
Low Moor:	41186
Bolton:	41103, 41104, 41189-41191 (5)
Kentish Town:	40930, 41050, 41051, 41054, 41071, 41077, 41117, 41199 (8)
Kettering:	41048, 41095
Leicester:	41089
Bedford:	41070, 41091, 41094 (3)
Nottingham:	40929, 41082, 41096 (3)
Derby:	40927, 41057, 41059, 41060, 41075, 41083, 41084, 41088, 41196, 41197 (10)
Sheffield Millhouses:	41062, 41063, 41072, 41079, 41198 (5)
Trafford Park:	40910, 41052, 41055, 41066, 41076, 41154, 41156, 41181 (8)
Leeds Holbeck:	40932, 41065, 41068, 41069, 41087, 41144 (6)
Manningham:	41067, 41137
Saltley:	40928, 41046
Bourneville:	40917, 40934, 41061, 41064, 41073 (5)
Bristol:	40935
Gloucester:	41047, 41058, 41074, 41078, 41097 (5)

Trafford Park's 1066 leaving Disley Tunnel and approaching New Mills with a Manchester-Derby local train, 13 June 1946. This engine has a strengthened frame extension. (MLS Collection)

It will be seen from the above that the largest single route concentration was on the North Wales coast and that with sixty, Scotland still had a significant number. The Midland Division had a similar number (61) but it must be remembered that the entire class of 45 ex-Midland Railway Compounds was based there too. When withdrawals began it was the Scottish engines that were the first victims and those based in Scotland were withdrawn early or were transferred to English sheds in the early 1950s.

Records of the running of the Compounds after nationalisation are rare as they were seldom seen on trains where they were extended. In 1950, M.N. Bland recorded a stopping train from Clapham Junction to Lancaster that was unusually made up of nine coaches weighing 290 tons gross and, equally unusually on this route, was hauled by two Compounds, 41080 and 41144. With no intermediate section scheduled for longer than seven minutes, the two engines cut the schedule by over half a minute for each, attaining around 50mph between stops, and a maximum of 62 between Hornby and Caton. They were used regularly, together with LMS 2P 4-4-0s, on the CLC Manchester-Liverpool and Manchester – Chester services, although it was noted in a copy of *Trains Illustrated* in the early 1950s the ex-GC D10 and D11 4-4-0s had the better acceleration from the many stops and an ex-GE D16/3, 62535, was also performing very well in comparison.

Holbeck's 1144 on the 3.55pm Manchester-Chinley local train passing Gowholes, 30 August 1947. (J.D. Darby/MLS Collection)

Llandudno Junction's 936 with the unique high-sided Compound tender after hauling the 2.50pm Llandudno to Birmingham on the Whitsun Bank Holiday as far as Crewe, 31 May 1947. It has now reacquired a smaller Fowler chimney (see earlier photo on page 216). (H.D. Bowtell/MLS Collection)

No 1048 rounds the curve into Clapham Junction past a busy scene in the goods yard with a Morecambe-Leeds train, 13 June 1947. (H.C. Casserley/MLS Collection)

No 1144 at Strines between Hazel Grove and New Mills with a Manchester-Sheffield stopping train, 7 August 1948. (R.D. Pollard/MLS Collection)

No 41076 of Trafford Park at Strines with a Derby-Manchester slow train, 18 June 1949. (MLS Collection)

Stirling's 40913 leaving Inverness with the 2.05pm to Forres and Aviemore, 11 October 1948. (J.D. Darby/ MLS Collerction)

Gloucester's 41074, which has obtained a tall chimney, runs into Blackwell station with a Birmingham-Gloucester stopping train, 1949. The only other engine noted with a tall chimney, apart from the Midland engines and 1045-1064 when first built, was Crewe's 41167 in the early 1950s. (G.A. Coltas/ MLS Collection)

Mr Twibell was a regular passenger in 1950 between Bedford and St Pancras on a morning commuter train that normally loaded to eight vehicles, 250 tons gross and was regularly hauled by a Compound. The train stopped at Flitwick and Luton only and was allowed 16 minutes for the 9.4 uphill miles, to Flitwick, another sixteen for the ten miles on to Luton and 36 minutes for the 30.1 miles on to London. The outline performance of five typical runs are tabled below:

	Bedford-St Pancras, 1950 8.06am Bedford				
	16 February 1950	1950	5.1950	27 May 1950	8 July 1950
	41075	41088	41006	41091	41049
	240 tons	250 tons	240 tons	181 tons	250 tons
Bedford – Flitwick	17m 14s	18m 07s	15m 49s	16m 14s	16m 05s
Speed at Ampthill	34 ½	36½	38	37½	40½
Flitwick – Luton	17m 08s	17m 04s	15m 17s	14m 42s	16m 21s
Speed at Leagrave	49 ½	46½	50	50½	49½

	Bedford-St Pancras, 1950 8.06am Bedford				
	16 February 1950	1950	5.1950	27 May 1950	8 July 1950
	41075	41088	41006	41091	41049
	240 tons	250 tons	240 tons	181 tons	250 tons
Luton – St Pancras	43m 14s	35m 56s	35m 25s	38m 04s*sigs	35m 41s
Speed at Radlett	54½	69½	74	68½	74
Speed at Elstree	42	51	50½	48	55
Speed at Hendon	56½	62½	64½	64½	67½

The first run was poor and got increasingly so with signs of poor steaming. One of the two best was with the 1903 built 41006 – several of these spent their last years at Bedford for such services. I've found a down log of the 2.07pm St Pancras-Leicester semi-fast in 1955 but the recorder only travelled as far as St Albans. No 41199 had seven coaches for 220 tons gross and departed 1½ minutes late via the slow line. It climbed the 1 in 178 out to Finchley Road, 3.5 miles, in 7 minutes 33 seconds topped at 37mph, touched 52 in the slight dip before Hendon and then climbed the five-mile 1 in 176 to Elstree at a minimum of 42mph, with a maximum of 62 on the 1 in 200 descent to Radlett. The three-mile 1 in 176 to St Albans brought 41199's train down to 45mph before the brakes were applied for the stop there, reached in 28 minutes 32 seconds having regained the 1½ minute late departure. Rugby's 41105 was also timed on an outer suburban train from Euston in August 1957, the same period when I was using the corresponding up working. It had eight coaches, 285 tons gross and experienced some difficulty leaving Harrow and Wealdstone where it was crossed to the slow line and dropped five minutes on the tight seven-minute schedule for the 4.6 miles to Bushey. It then held time to Watford but again slipped badly, leaving the station as far as Watford Tunnel and dropped two minutes to King's Langley. It recovered after that and gained a few seconds on the runs to Apsley and Hemel Hempstead where the recorder alighted. I have to say that my experience on the up Rugby semi-fast due off Willesden at 6.09pm on the 5.5 mile run to Euston was fairly pedestrian, the ex-store Compounds that summer (41093, 41105, 41113, 41122, 41162 and 41172) averaging ten minutes late arriving into Willesden. Nos 41093 and 41162 looked to be in the best condition and the only punctual arrival that I can remember was made by 41122. The load was fairly substantial for Compounds on a stopping service – seven or eight coaches and a couple of parcel vans. The train was diagrammed for a Rugby 'Black 5'.

Rugby's 41113 was transferred to Lancaster in May 1958 and was timed by Mr Bland on the lightweight four-coach 120-ton 5.10pm Bradford Forster Square-Morecambe on the 29th of the month. It left Shipley on time and worked up to 62mph before Keighley where it was checked to 30mph by 'Black 5' 44854 on the 4.55pm Leeds-Morecambe. It then accelerated to 61 down the short 1 in 247 to Steeton and carried on accelerating on the level through Cononley to a full 70mph, arriving at Skipton 1¼ minutes early. Bland returned to Leeds on the 4.20pm from Morecambe which he joined at Skipton with 41101 and seven coaches, 212 tons, leaving seven minutes late and after touching 60 at Steeton, dropped three coaches off at Keighley. It cantered on with 57 before Bingley and 55 after, passing Leeds Junction just under six minutes late, then speed rose to 67mph at Calverley and, easing after Kirkstall Junction, reached Leeds City five minutes late.

My final records are of the performances of Compounds on *Trains Illustrated* rail enthusiast excursions. On 29 September 1955, 46100 *Royal Scot* worked a special to Derby where Compounds 40927 and 41167 took over for the run to Buxton. Unfortunately, we only have a record of the return downhill journey. The pair took the 11-coach 360 ton gross train down through Bakewell at just under 60mph before a signal stop at Rowsley and then a stop at the platform to pick up a pilotman. The pair then accelerated the train to 66mph at Belper, but was 11 minutes late into Derby, where 46100 took over again, departing 15 minutes late. Then nearly three years later 41100 (Leeds Holbeck) and 41063 (Bradford Manningham) took over at Leeds from A1 60157 *Great Eastern* that had worked the *Pennine Limited* from King's Cross. The run was timed by Mr B.I.Nathan.

Leeds City-Crewe, *Pennine Limited*, **26 April 1958**
41100- 55A & 41063 - 55F
9 chs, 320/340 tons (incl *Devon Belle* observation coach)

Miles	Location	Times	Speeds		Gradients
0	Leeds City	00.00		1¼ L	
2	Farnley Junction	05.41	33/30	1 L	1/145 R
4.8	Morley	11.25	36	1¾ L	1/120 R
6.8	Batley	16.06	44/sigs		1/138 F
9.2	Dewsbury	18.40	35*		1/143 F
10.8	Thornhill Junction	19.18	46	½ L	
12.6	Mirfield	23.44	44½	T	
13.6	Heaton Lodge Junction	25.27	41/45	¼ E	
<u>17.2</u>	<u>Huddersfield</u>	<u>31.50</u>		T	
0		00.00		1½ L	
3	Golcar	07.19	34		1/105 R
4.4	Slaithwaite	10.03	35/sigs		1/105 R
7.3	Marsden	15.11	24*/39	1½ L	1/105 R
	Diggle Tunnel	-	41	L	
<u>10.8</u>	<u>Diggle</u>	<u>22.08</u>		½ E	
0		00.00		¾ L	
1.2	Saddleworth	02.42	46		
2.8	Greenfield	04.12	51	1 L	1/125 F
4.6	Mossley	07.04	51/49*		
<u>7</u>	<u>Stalybridge</u>	<u>11.14</u>		<u>1 L</u>	
0		00.00		¾ E	
	Denton Junction	05.52			
	Heaton Norris Junction	11.45			
	<u>Stockport</u>	<u>14.14</u>		1½ E	
0		00.00		2 E	
2.2	Cheadle Hulme	04.49	44/pws 30*		
4.5	Handforth	09.04	35		1/445 R
5.9	Wilmslow	12.42	32*		
7.6	Alderley Edge	14.30	58		L
10.7	Chelford	17.55	60		L
14.6	Goostrey	21.53	pws 40*		
16.7	Holmes Chapel	24.29			1/335 F
20.3	Sandbach	31.04	easy		
<u>25</u>	<u>Crewe</u>	<u>41.04</u>		<u>2 E</u>	

The schedule was relatively easy, which the Compounds were able to achieve without too much effort apart from the climb to Diggle Tunnel and on the last stretch from Stockport, when 41063's fire was clinkering badly and most of the work was performed by the pilot, 41100. No 41123 of Gloucester was used as the train engine for the May 1958 'Talyllyn AGM' special from Paddington to Shrewsbury and back, a pair of GW 'Dukedogs' taking over for the Cambrian section to Towyn.

As an example of the use of Compounds in their last days, 41157 was piloting 'Jubilee' 45602 *British Honduras* on the 12-coach 7.05am from Sheffield to St Pancras on 10 June 1959. The pair left Derby on time and got to Leicester (30 miles) in 32 minutes 31 seconds, top speed 64mph. The 'Jubilee' and Compound then showed a bit more energy in climbing the 1 in 161 to Kibworth at 52mph and the long 1 in 132 to Desborough at 43. After that, things deteriorated, with signal checks at Glendon and Rushton, a p-way slack at Kettering Junction and after a reasonable climb to Sharnbrook the pair clearly coasted down the bank only averaging 50mph from the summit to passing Bedford Junction, a sign that one or both were short of steam. The pair averaged 51mph over the long 1 in 200 to Leagrave, passing Luton 10 minutes late. Then followed another p-way slack at Harpenden, St Albans was passed 13 minutes late, and the pair only averaged 58mph over the thirteen mile downhill section to Hendon (interrupted by two miles of 1 in 200 up to Elstree). Despite an unchecked run into the terminus, the train was 14 minutes late, having taken a few seconds under 113 minutes on the fast 99 minute schedule.

Thirteen Compounds were withdrawn in 1959, leaving just six remaining. In 1960, four of these were withdrawn – 41157 from Derby in May, 41162 from Rugby in June, 40907 from Sheffield Millhouses and 41063 from Bradford Manningham in September. The last two survived until 1961, 40936 from Monument Lane in January and 41168 from the same depot in July.

The large tendered Compound, 40936, pauses at Stafford with a Blackpool-Birmingham return excursion, 5 July 1952. Note the low position of the BR emblem on the tender but lined up with the cab number. (MLS Collection)

Corkerhill's 40906 at Ayr with a Kilmarnock local train, 27 July 1951 (E.R. Morten/MLS Collection)

Accrington's 41185 assists a 'Black 5' to the summit at Diggle passing Linthwaite on a Newcastle-Liverpool express, c1953. (MLS Collection)

The LMS 4P Compound • 257

Trafford Park's 41052 at Buxworth Junction on the 10am Manchester-Derby, 28 June 1952. (B.K.B. Green/MLS Collection)

Gloucester's 41097 ascending the Lickey bank pushed by a 'Jinty' tank with a Bristol-Birmingham stopping train, c1953. The frame extension is very clear in this photograph. (MLS Collection)

Holbeck's 41100 at Shipley with a railfans' excursion from Bradford to Derby Works, 6 September 1953.
(A.C. Gilbert/MLS Collection)

No 41113 of Longsight with a light stopping train from Crewe to Manchester near Goostrey, 1951.
(MLS Collection)

No 41140 of Kingmoor assists a 4F 0-6-0 43896 with a northbound goods train near Lazonby, 22 May 1951. (R. Hewitt/MLS Collection)

No 41154 of Trafford Park stands at Marple station with the 1.04pm Manchester-Sheffield stopping train, 25 May 1953. (B.K.B. Green/MLS Collection)

Another Trafford Park Compound, 41159, on the same 1.04pm Manchester-Sheffield local, overtaking 4F 43963, at Chinley, May 1953. (B.K.B. Green/MLS Collection)

A third view of the 1.04pm Manchester-Sheffield with another Compound, 41123, at Didsbury, 31 May 1954. (A.C. Gilbert/MLS Collection)

Crewe's 41119 pilots an unidentified 'Jubilee' on a southbound West Coast express at Carlisle Citadel station while LMS 2P 40565 of Kingmoor is on station pilot duty, c1955. (MLS Collection)

No 41077 of Kentish Town at St Pancras with a suburban commuter train to Bedford, 26 June 1954. (J.H. Aston/ MLS Collection)

Trafford Park's 41066 on arrival at Aintree with a Grand National special excursion, flanked by a 'Black 5' 45006 and B1 61265 of Sheffield which have arrived earlier, 24 March 1956. (MLS Collection)

Derby's 41050, with six months to go before withdrawal, oozes steam from all parts while traversing the snows of the Peak District near Chinley, 14 February 1956. (MLS Collection)

No 41066 at Altrincham with a Manchester-London fitted freight, 20 June 1957. No 41066 had a year to go before withdrawal from Saltley – it was transferred there just a couple of months after this photograph was taken. (H.D. Bowtell/MLS Collection)

No 41102 of Blackpool polished up for the *Northern Dales Rail Tour,* jointly organised by the Stephenson Locomotive Society and the Manchester Locomotive Society, at Manchester Victoria, 4 September 1955. The rail tour started from Manchester at 9.23am and the Compound hauled the train to Tebay via Blackburn, Hellifield and Clapham Junction, where the tour was taken on by ex North Eastern engines to Darlington and Northallerton, the Compound bringing the train back from Garsdale in the evening. (MLS Collection)

No 41100 and 41063 recover from a signal check to 24mph on the climb to Diggle at Marsden with the 'Trains Illustrated' sponsored *Pennine Limited* rail tour, 26 April 1958. (A.C. Gilbert/MLS Collection)

No 41101 at Blackpool with the return 'Andy Capp' *Daily Mirror* (wet) Bank Holiday excursion from Manchester to Blackpool, 3 August 1959. The repainting in the garish red and yellow livery specified by the newspaper company was carried out at Gorton Works (it is rumoured the staff at Derby Works refused to carry out the painting which they labelled 'sacrilege'!). It was fitted with a chime whistle from a 'Clan' pacific awaiting overhaul at the Works. (MLS Collection)

Chapter 7
THE LMS 2P

Design & construction

During the early years of the LMS Company, the focus of the motive power authorities was on the provision of suitable locomotives for main line express work, culminating eventually in the decision to acquire the three-cylinder 'Royal Scot' class. In the meantime, the secondary services had soldiered on with increasingly aged 4-4-0s, the most modern being the superheated class 2s of the former Midland Railway – some theoretically forty years old or more, but 'replaced' in the previous decade with new boilers and cylinders and in many cases, new frames as well. The former LNWR had the 'Precursor' and 'George V' 4-4-0s as well as the 'Experiment' and 'Prince of Wales' 4-6-0s, now largely displaced from express work, but many of them were incurring costly repairs after years of working to their limit on heavy West Coast services. The Caledonian Railway had provided some 4-4-0s of relatively modern vintage – the McIntosh Dunalastairs and the Pickersgill developments of them, but the other Scottish 4-4-0s of the former G&SWR and Highland railways were life-expired, as were the smaller engines of the Lancashire and Yorkshire Railway. The traffic Department demanded replacements that were less costly.

The motive power authorities were still largely from the Midland Railway background and rather than produce a new design incorporating the experience of the other railways, particularly the GW's use of outside cylindered 4-6-0s for mixed traffic work or Maunsell's rebuilding of the SE&CR 'D' and 'E' 4-4-0s with new superheated boilers and long travel valves, the 'quick fix' was to update the Midland superheated '483' class, but still with short travel valves. Initially, an improved version with outside cylinders and Stephenson's valve gear was considered but rejected as such a machine on some secondary routes would face gauging problems. In 1928, therefore, Derby works started turning out 2P 4-4-0s of basically the Midland design, but with 6ft 9in coupled wheels instead of 7ft, and cut down boiler mountings. The dimensions and comparison with the Midland '483' class were as follows:

	LMS 2P	Midland '483'
Cylinders	19in x 26in	20½in x 26in
Coupled wheel diameter	6ft 9in	7ft 0in
Bogie wheel diameter	3ft 6 ½ in	3ft 6½ in
Boiler pressure	180lbs psi	160lbs psi
Heating surface	1,410sq ft	1,410sq ft
Grate area	21.1sq ft	21.1sq ft
Axleload	17¾ tons	17½ tons
Weight		
– Engine	54 tons 1 cwt	53 tons 7 cwt
– Tender	41 tons 4 cwt	41 tons 4 cwt
– Total	95 tons 5 cwt	94 tons 11 cwt
Water capacity	3,500 gallons	3,500 gallons
Coal capacity	4 tons	4 tons

In view of the help Fowler had received from Richard Maunsell of the Southern in acquiring the drawings of the 'Lord Nelson' 4-6-0 as a model for the 'Royal Scots', it is surprising that he did not ask for similar help for the 4-4-0 as Maunsell's 'E1' and 'D1' rebuilds of 1919 of very similar dimensions were gaining a good reputation. The Southern 4-4-0s had 6ft 6in coupled wheels but the main difference that transformed their performance was the use of an efficient front end design with long travel valves in the Churchward tradition.

Fifty engines of the design were constructed at Derby in 1928 and numbered 563-612, showing their ancestry by following on immediately from the numbers of the '483' class. However, 575, 576 and 580 were transferred to the Somerset & Dorset Railway becoming their 44-46 (see Chapter 2, page 116) and were replaced by three extra in 1929, and given the 575, 576 and 580 numbers. Similarly, after early complaints of poor steaming with the new engines, 572 was experimentally fitted with Owen's double port exhaust valves and was renumbered 601 before entering service, a replacement 572 being immediately constructed. Nos 613-628 followed in 1929 and 629-632 in 1930 when the Somerset & Dorset engines returned to LMS operation and became 633-635. The class was continued in 1931 and 1932 with two batches, 636-660 and 686-700 being built at Crewe, while 661-685 were constructed at Derby. Then, with the arrival of Stanier from Swindon, further possible orders were terminated. There was one minor attempt at improvement in 1933 when 633 (former S&D 44) and 653 were fitted with Dabeg feedwater heaters, previously fitted to withdrawn ex-Midland class 3 'Belpaires', but the extra efficiency did not offset the extra maintenance costs and despite 5 per cent coal and 8 per cent water savings, no further engines were so equipped.

Despite the limitations of the design and the missed opportunity to make them better, they were economical machines, light of repair as well as running costs, although much of their work was

The first LMS 2P, 563, built in 1928, in Works grey. (Railway Photographs/MLS Collection)

No 564 under construction at Derby in 1928. (H.C. Casserley/MLS Collection)

No 572 renumbered 601 before entering service, experimentally equipped with Owen double port exhaust valves, Derby, 1928. (MLS Collection)

No 576, built at Derby in 1928 and transferred to the Somerset & Dorset Railway as No 45, seen here at Bath depot in 1929. It was taken back by the LMS in 1930 and given the number 634.
(Locomotive & General/ MLS Collection)

Ex Somerset & Dorset Railway No 44, renumbered 633 in 1930 and fitted with a Dabeg feedwater heater system off Midland class 3 'Belpaire' 702 in 1933.
(F. Moore/MLS Collection)

Ex Somerset & Dorset Railway No 46, taken into LMS stock in 1930, repainted and renumbered 635. (MLS Collection)

light, either on local three and four coach stopping trains or piloting expresses. In fact, their low average repair costs were the best of any LMS passenger locomotive. They were welcomed in Scotland in particular, especially on the former G&SWR lines as they were such a considerable improvement on the Manson engines still operating there.

Two – 591 and 639 – were withdrawn in 1934 after being heavily damaged in a collision at Port Eglinton Junction, Glasgow, but the rest were taken into BR ownership in 1948 and renumbered 40563-40700. The first being built in 1928 had just missed out on the LMS red livery, so initially they were black with red lining, then plain black, and most in BR days received the mixed traffic lined black livery. The first withdrawal for other than accident damage was 40676 in 1957, but as much of their work was then being covered by diesel multiple units introduced in the mid-1950s, withdrawals began in earnest in 1959, with forty-three being taken out of traffic that year. Ten more went in 1960 and then the majority of the rest – sixty-six – went in 1961. The last fifteen were withdrawn in 1962. The last survivor was 40670, condemned in November 1962, only two months after the last pair of the Midland engines, 40453 and 40537. None of the LMS 2Ps achieved anywhere near a million miles in traffic, the 1928 built 633 which operated over the S&D system for its first five years reaching 789,784 and the 1931 653 just 633,734. These were the two Dabeg feedwater heater fitted locomotives whose performance was watched more closely, but their mileages were typical of other engines of the class.

No 591 at Glasgow St Enoch on Easter Monday 1931. It was involved in the collision with 639 at Eglinton Street, Glasgow, in 1934 and cut up as beyond economic repair.
(G.A. Coltas/MLS Collection)

No 663 built at Derby in 1931 and allocated to Scotland, c1935.
(J. Lord/MLS Collection)

No 653 built at Crewe in 1931 and fitted with Dabeg feedwater heater off Midland 'Belpaire' 706 in 1933, at Crewe North shed, 19 April 1936.
(W.A. Camwell/MLS Collection)

No 697 built at Crewe in 1932 and fitted with a stovepipe chimney as part of the experiment to improve steaming. No others were converted. It is seen at Saltley, 17 March 1935.
(W.A. Camwell/MLS Collection)

No 700, the last of the class, built at Crewe in 1932, at Saltley, c1938. The Crewe engines had a rougher finish than the Derby built engines and the rivet heads are clear in this photograph.
(MLS Collection)

No 40613 in first BR guise, unlined black and still awaiting new BR smokebox numberplate, at Carlisle Kingmoor depot, 17 October 1949.
(J.D. Darby/MLS Collection)

No 40633, one of the two Dabeg water heater fitted engines, still with its equipment in place at Burton-on-Trent shed, 2 June 1950. Burton's 4F 43837 is in the background. (H.C. Casserley/MLS Collection)

No 40568 is one of the 1928 first series of LMS 2Ps that were allocated for work on the Somerset & Dorset but on loan retaining LMS numbers. It is photographed on Bath shed, in BR mixed traffic lined livery, 2 September 1951. The tablet catcher for single line operation is seen on the front of the tender. (G. Harrop/MLS Collection)

No 40677 under repair at Rugby shed, seen during the visit of the Charterhouse Railway Society I led on 17 May 1956. (David Maidment)

No 40578 was one of the 2Ps painted in typical style of Scottish-based engines with large cabside numerals. It is at Glasgow St Enoch, 26 April 1952. (A.G. Ellis/MLS Collection)

No 40613, recently overhauled and repainted with the large final BR tender emblem, but in store at Carlisle Kingmoor, c1960. It was withdrawn in 1961. (MLS Collection)

Operations

The most urgent need for the new 2Ps was in Scotland, particularly the former Glasgow & South Western services from St Enoch to Kilmarnock and the Ayrshire Coast. Over half of the new class were allocated there from the beginning. Some of the first also went to assist on the Somerset & Dorset with three that were actually transferred to the ownership of that company. The allocation after delivery of all 138 engines was:

Midland Division (former MR):	563-569, 600-602, 606-612, 628-632, 696-700 (27)
Central Division (former L&Y):	580-589, 676-685, 690, 691 (22)
Western Division (former LNWR):	651-660, 671-675, 692-695 (19)
Northern Division (Scotland):	570-579, 590-599, 603-605, 613-627, 636-650, 661-670, 686-689 (67)
Somerset & Dorset:	44-46 (633-635 from 1930)

The first area that attracted the attention of the train timers to the LMS 2Ps was the Bath-Bristol route of the Somerset & Dorset, especially between Bath and Evercreech over the Mendips at Masbury. Southbound from Bath, after the first level half mile, there are two miles of 1 in 50 to Coombe Down Tunnel, a short dip to Midford, most at 1 in 100, the six miles of undulating track to Radstock, then eight miles of stiff climbing at mostly 1 in 50 to Masbury summit, easing briefly at Binegar, before the final mile of 1 in 73. Then follows three miles of steep descent to Shepton Mallet. The *Railway Magazine* published a number of runs in 1936 timed by D.S.Barrie. All were on the *Pines Express*, the only named train entrusted to a 2P unassisted.

Bath-Evercreech Junction, *Pines Express*, 1935.

Miles	Location	630 5 chs, 161/170 t				697 6 chs, 179/190 t 17 June 35				631 7 chs, 201/210 t				630 6 chs, 190/200 t 3 December 35			
		Times	Speeds			Times	Speeds			Times	Speeds			Times	Speeds		
0	Bath	00.00				00.00				00.00				00.00			
0.5	Bath Junction	01.51				01.56				01.54				01.58	25		
2.5	Coombe Down	07.48	18½			07.57	20			08.20	12½			08.22	15		
4.3	Midford	10.26	60		½ L	10.55	52		1 L	11.13	57		1¼ L	11.10	55		1¼ L
6.8	Wellow	13.19				14.15	47			14.34				14.19	43/51		
10.7	Radstock	18.09	23*		¾ E	19.13	52/46*		¼ L	19.02			T	19.27	21*		
12.5	Midsomer N	23.48				23.52	20			22.47	15			24.39	19		
14.5	Chilcompton	29.55	17½			29.50	21			30.02	20/17			31.00	21		
17.1	Binegar	35.40	37			35.29	37			36.08	38½			36.40	37		
18.7	Masbury	38.21	30		¾ E	38.04	30		1 E	39.00	32½		T	39.25	29		½ L
21.9	Shepton Mallet	42.14	60		¾ E	41.55			2 E	42.48	64		¼ E	43.03	56		T
24.9	Evercreech N	45.48				04.53	62			46.10				46.18	65		
26.4	Evercreech Jn	48.27			1½ E	07.05				48.41			1¼ E	48.41			1¼ E

In the same year, 'Black 5' 5440 made similar times with 8 coaches, 270 tons, and 5432 with 2P 700 as pilot took 11 coaches, 365 tons in a similar time also.

The 2Ps were used on Leeds-Heysham boat trains, and 660 was timed in July 1936 on the 8.35pm Leeds connection to the night boat to the Isle of Man. It started with seven coaches, made up to nine at Shipley, 220 tons gross, and was nine minutes late away from that point. 62mph was reached after Keighley, and an excellent climb was made from Hellifield to Giggleswick, 56 falling to 47mph. After Clapham Junction a maximum of 72mph was attained and arrival at Lancaster Green Ayre was seven minutes late.

Nos 606-612 were transferred from the Midland Division to Scotland in 1930, bringing the Scottish fleet up to 74 locomotives of the class. The engines were used on a number of secondary services – commuter trains in and out of Glasgow St Enoch, semi-fast services to the Ayrshire Coast, local stopping services throughout the former G&SWR system and the rest of the LMS and piloting expresses hauled by Fowler 'Royal Scots' and Compounds, and later by 'Black 5s' and 'Jubilees' on both the Midland and West Coast main lines.

Nos 591 and 639 were scrapped after the Eglinton Street Junction collision in 1934 and were replaced by 566, 600 and 602 from the Midland Division and the Western Division acquired 628, 629 and 635 from the Midland also. The allocation then until nationalisation remained stable at:

Midland Division (incl S&D):	17
Central Division:	22
Western Division:	22
Northern Division (Scotland):	75

On the Midland Division the LMS built engines were used indiscriminately with the many Midland 2Ps ('483' class). The most specialist use was on the Bath-Bournemouth route of the S&D, being based at both Bath Green Park and Templecombe and used to pilot through trains like the *Pines Express* as well as hauling the three coach stopping trains within the S&D system. On the Midland main line,

No 665, built in 1931 at Derby and allocated immediately to the Northern Division at Kilmarnock, with a stopping train to Ayr, with Whitelegg 4-6-4T 14503 alongside, c1932. (H.N. James/MLS Collection)

No 698, built at Crewe in 1932, brings a Manchester-Derby-Nottingham train through Rowsley station, 26 May 1934. (MLS Collection)

No 635, built in 1928 for the Somerset & Dorset Railway, numbered 46, and absorbed into LMS stock in 1930, at Bournemouth West with an afternoon stopping train for Templecombe, 1934. (MLS Collection)

they were used on secondary services in the East Midlands around Derby, Nottingham and Leicester as well as piloting main line services whose loads exceeded the limits laid down by the LMS operators, a practice that continued long after the provision of Stanier 4-6-0s on the route.

On the Central Division, they were allocated to the following depots – Bank Hall, Southport, Newton Heath, Bury, Accrington, Bacup, Darwen, Low Moor, Wakefield and Goole. Their main use was on local services around East Lancashire – Preston-Rochdale, Chorley, etc – as well as to Liverpool and Blackpool. Few took an interest in their running at this time but alongside is the log I've found of a run between Manchester and Liverpool, albeit on a lightly loaded train that did not need much strenuous effort though it demonstrates free running and a preparedness to run at high speed .

Manchester Victoria-Liverpool Exchange, 16 July 1931
588
4 chs, 110 tons

Miles	Location	Times	Speeds	Gradients
0	Manchester Victoria	00.00		
0.7	Salford Central	02.10		
2.3	Pendleton	04.25	47/60	L
4.4	Pendlebury	07.15	40	1/99 R, 1/80 R
7.5	Walkden	10.40	60	
11	Atherton	13.50	69	1/150 F
12.7	Daisy Hill	15.25	59/67	1/150 R, 1/265 F
15.2	Hindley	17.50	74/ pws 45*	1/90 F
18	Westwood Park	20.20	70	1/154 F
19.3	Pemberton	22.10		1/92 R
21	Orrell	24.25	40	1/92 R
24.4	Rainford	28.05	72/77	1/118 F
29.5	Kirkby	32.20	72	1/247 F
31.4	Fazakerley	34.00	69	L
35	Sandhills	37.40	62	
36.5	Liverpool Exchange	40.30		

Crewe built 651 leaving Llandudno Junction with a stopping train for Chester, 21 June 1936.
(E.R. Morten/MLS Collection)

Another Crewe
1931 built 2P, 658, at Prestatyn with a Manchester-Llandudno bank holiday relief, 23 May 1936. (E.R. Morten/MLS Collection)

695, built at Crewe in 1932, picks up water from Dillicar troughs ready to climb to Shap summit with a train of milk tank empties, c1934. (MLS Collection)

On the Western Division they were scattered, though concentrations went to the coastal outposts such as the North Wales and Cumberland sections. The Rugby-Nuneaton area had a few and later more went to the Preston and Stafford areas for local services. Carlisle, Preston and Crewe North based 2Ps were used frequently for piloting main West Coast expresses, assisting all classes except the Stanier pacifics. They were also often used on engineer and general management saloon inspection duties, 672 and later 40646 being allocated to Watford for the purpose.

The archive of the Railway Performance Society has unearthed one pre-war 2P unassisted run timed by D.S.Barrie on the southern end of the West Coast line, a Friday evening relief train to Northampton in the summer of 1937. Although the 2Ps were not reputed to be outstanding performers, with a comparatively light load they could be speedy and on this occasion the climb to Tring was excellent.

Euston-Northampton, 30 July 1937
4.15pm Euston
672 - Watford
7 chs, 198/210 tons

Miles	Location	Times	Speeds	
0	Euston	00.00		
1	Camden	03.47		
3	Kilburn	07.03	53/sigs 30*	
5.4	Willesden Jn	10.12	45	¼ L
11.4	Harrow	16.57	56	
13.3	Hatch End	18.55	60	
17.4	Watford	22.54	68	1 E
24.5	Boxmoor	29.27	63½	
31.7	Tring	36.30	60	2½ E
36.1	Cheddington	40.15	78	
40.2	Leighton	43.25	76	
45	MP 45	47.19	72/sigs 10*	
46.7	Bletchley	50.40		2¼ E
0		00.00		T
5.7	Wolverton	08.08	65	
10.7	Hanslope	12.55	60	
13.2	Roade	16.40	sigs 15*	1¾ L
15.9	Middleton	20.34	67	
19.1	Northampton	24.30		½ L

In Scotland the former G&SWR routes had the majority:

Hurlford:	22
Corkerhill:	10
Ayr:	4
Greenock:	2
Girvan:	2
Dumfries:	9
Stranraer:	3
Ardrossan:	12

Many logs by D.S. Barrie on the ex-G&SW lines in 1937 and 1938 exist. No 605 on a Stranraer day boat train with 200 tons gross in July 1937 ran the 21 miles from Newton Stewart to New Galloway in 35 minutes, a minute less than schedule, holding 25mph on the eight miles of 1 in 80 to Gatehouse of Fleet, with maxima of 58 before and after. The train had left Stranraer nearly 25 minutes late awaiting the boat and had recovered six minutes of the lost time by Dumfries. Barrie commented that the engine was being worked 'very hard'. The same engine on the 4.2pm Carlisle-Stranraer Harbour in August 1938 with eight coaches, 280 tons gross, ran the 33.1 miles from Carlisle to Dumfries in 41¼ minutes, including a signal check to walking pace before Kingmoor and a p-way slack to 35mph after Ruthwell, dropping 2¼ minutes, maximum speeds, 60 at Eastriggs and 64mph at Racks. On a stormy August night in 1938, the 9.15pm Glasgow St Enoch-St Pancras sleeper train left with 2P 649 assisting 'Jubilee' 5594 *Bhopal* on twelve coaches, 405 tons gross, as far as New Cumnock. The train stopped at Kilmarnock and the pair accelerated hard over the first couple of miles to

Hurlford 'passed at 53mph' and fell to 36 after the five mile climb at 1 in 100 to Garrochburn. They recovered to 60mph at Mauchline, passed on time in 14 minutes from Kilmarnock, fell to 48 on the 1 in 180 to Auchinleck, 55 at Old Cumnock and finally a minimum of 46 on the 1 in 145 past Polquhap Sidings. The 21.1 miles to the New Cumnock stop was completed in ten seconds less than the 30 minutes allowed.

5582 *Central Provinces* had a heavy 13-coach load, 415 tons gross on the overnight London from Carlisle to Glasgow St Enoch and was provided with 604 to assist. Schedule to Kilmarnock was just kept despite a p-way slack after Old Cumnock and a signal check outside Kilmarnock station (64 minutes for 58 miles). Speed ranged between 52 and 57mph up the long 1 in 200/1 in 150 to Thornhill, fell to 49 at Carronbridge at the summit of the 1 in 150 and achieved 72 after the 20mph p-way slack and a full 80 down the 1 in 99 before Hurlford. Signal checks on the last section made the train two minutes late at St Enoch. Mr Barrie continued to travel in the area during the war years and found 616 assisting 5617 *Mauritius* on a massive 500 ton load on the night boat train from Stranraer Harbour in April 1944. Despite the load, the train was eight minutes early arriving at Dumfries taking just one minute over two hours for the 73.7 heavily graded and speed restricted miles.

After nationalisation, there were minor changes but in 1950 the allocation of the fleet by former Motive Power Division and depot was:

Midland Division
Templecombe:	40563, 40564, 40634
Bath:	40568, 40569, 40601, 40679, 40696-40698, 40700 (8)
Saltley:	40581, 40582, 40584, 40585
Manningham:	40567
Normanton:	40630
Derby:	40632
Burton:	40633
Buxton:	40655, 40692

Most of their post-war work on the former Midland Railway lines was of a secondary nature. Examples I've found include the Burton based Dabeg feedwater heater fitted 40633 which worked the 12.40pm Leicester-Worcester south of Birmingham New Street on 31 July 1948, a 200 ton train that was normally hauled by a 'Black 5'. No 40691 of Hasland was power for the 5.45pm Nottingham-Sheffield in April 1956 which ran punctually throughout but with a featherweight load of four coaches. In April 1954, a pair of the Bath engines took over a *Trains Illustrated* special from 'Schools' 30932 *Blundells* at Bournemouth and took it to Bath, then returned to Templecombe, where 30932 whisked the railtour back to Waterloo.

No 693 piloting 468 (a '2203' rebuilt as a '483' superheated class) at Chinley with a Manchester-Sheffield stopping train, the lack of cleanliness of both engines typical of the immediate post-war period, 30 August 1947. (MLS Collection)

Bath-Templecombe, 25 April 1954
Trains Illustrated 'Somerset & Dorset' Special
3.55pm Bath-Templecombe/Waterloo
40601 & 40698
8 chs, 262/290 tons

Miles	Location	Times	Speeds		Gradients
0	Bath	00.00		½ L	
2.5	Coombe Down	-	30/16		1/50 R
4.3	Midford	10.21	52	¾ L	1/100 F
6.8	Wellow	-	pws 15*		
10.7	Radstock	19.55	29	1 L	
12.5	Midsomer Norton	24.12			1/50 R
14.5	Chilcompton	30.30	18		1/50 R
17.1	Binegar	36.29	36		L, 1/173 R
18.7	Masbury	39.20	28/55	2¾ E	1/63 R
21.9	Shepton Mallet	42.50	63	4¼ E	1/50 F
26.4	Evercreech Junction	48.29	(46 net)	7½ E	

No 40698 was detached at Evercreech Junction, and 40601 took the train onto Templecombe, where 30932 took over for the run to Waterloo.

No 634 (formerly S&D 45) at Chilcompton with the 3.10pm Bath-Bournemouth, 3 October 1949. (J.D. Darby/MLS Collection)

No 40696 at Poole with a summer timetable Bournemouth West–Sheffield express, June 1951. (G.M. Shoults/MLS Collection)

No 40634 (S&D 45) at Templecombe, 18 April 1952. (J. Davenport/MLS Collection)

No 40563 pilots a BR 'Standard 5' near Radstock with a train for Bournemouth, c1955.
(N.K. Harrop/MLS Collection)

No 40601, an S&D regular, pilots Bath based BR 'Standard 5' 73049 nearing Masbury summit, 26 July 1958.
(MLS Collection)

A few minutes later Templecombe based 40563 assists Bulleid unrebuilt West Country 34099 *Lynmouth* with another holiday express for Bournemouth, 26 July 1958. (MLS Collection)

Bath's 40568 with a local train typically formed of four ex-SR Maunsell coaches, at Binegar, 26 July 1958. (MLS Collection)

Central Division (ex L&Y)
- Brunswick: 40583, 40683
- Southport: 40690
- Wigan: 40580, 40587, 40684
- Patricroft: 40628, 40635, 40678
- Accrington: 40676, 40677, 40680, 40681
- Bacup: 40682, 40691
- Lower Darwen: 40588
- Goole: 40586, 40589, 40685

As well as pottering round with two/three coach locals in East Lancashire, the 2Ps worked the former CLC Manchester-Chester line through the Delamere Forest and the semi-fast Manchester-Liverpool trains. No 40679 of Brunswick worked the 3.30pm Manchester on 13 March 1950 with 6 coaches, 162 tons, and ran punctually including a p-way slowing to 27mph before Farnworth, with maximum speeds of 59mph before Warrington and a few miles before Liverpool.

No 40685 working a Leeds-Morecambe relief train past Long Preston, 29 June 1957. (B.K.B. Green/MLS Collection)

No 677 leaves Chatburn with the 6.43pm local train to Bolton, 21 June 1947. (J.D. Darby/MLS Collection)

Accrington's 681 wanders round the industrial part of East Lancashire with a two-coach local train, 28 May 1947. (H.D. Bowtell/MLS Collection)

No 40679 newly transferred from Bath to Liverpool Brunswick, with a Liverpool-Manchester train at Glazebrook, 25 March 1950.
(MLS Collection)

Brunswick's 40683 crossing the Mersey with a Cheshire Lines Committee (CLC) local train, 22 March 1951.
(T. Lewis/MLS Collection)

Chester's 40658 leaving Delamere on the CLC with a Chester-Manchester local train, 9 August 1953.
(MLSV Collection)

Longsight's 40693 stands at Liverpool Lime Street with a stopping train for Southport, 1958.
(E. Oldham/MLS Collection)

Western Division (ex LNWR)
- Upperby: 40652, 40699
- Barrow: 40654
- Workington: 40656, 40694, 40695
- Preston: 40565, 40631
- Rhyl: 40629, 40646, 40671, 40675
- Chester: 40658
- Crewe North: 40659, 40660
- Longsight: 40674, 40693
- Northampton: 40653, 40657
- Watford: 40672

I expected to find some logs of these engines in the RPS archives, especially along the North Wales coast and possibly on double-headed West Coast expresses, but all I could find was the Crewe based Dabeg water-heater fitted 40653 on a Crewe-Chester train which managed to take a pedestrian 32¾ minutes to get to Chester.

Scotland
- Hurlford: 40566, 40570-40573, 40593, 40597, 40605, 40612, 40617-40619, 40643-40645, 40661-40663, 40665, 40666, 40686-40689 (24)
- Ayr: 40574, 40575, 40590, 40610, 40638, 40640, 40647, 40648, 40664, 40670 (10)
- Dumfries: 40576, 40577, 40614
- Ardrossan: 40578, 40579, 40606-40609, 40624-40626, 40667-40669 (12)
- Carstairs: 40592
- Corkerhill: 40594-40596, 40598, 40599, 40604, 40620, 40621, 40627, 40636, 40637, 40641, 40642, 40649, 40651 (15)
- Stranraer: 40600, 40611, 40616, 40623
- Kingmoor: 40602, 40613, 40615
- Carlisle Canal: 40673
- Kittybrewster: 40603, 40622, 40650

Most of the logs of 2P performance that I have found relate to their work in Scotland in the 1950s. Mr D.S. Barrie clearly spent some time in the area post-war and was assiduous in timing his runs. Corkerhill's 40651 hauled the 8.25am Glasgow-Ardrossan on 30 August 1951 with eight non-corridor coaches, and after a signal stand for fifteen seconds at Cork Street, arrived at Paisley Gilmour Street ¾ minutes late and left on time. It passed Dalry at 59mph (maximum speed), but was brought to a stand outside Ardrossan where it terminated 3½ minutes late. Ardrossan's 40578 hauled the four-coach 1.40pm Ardrossan-Glasgow St Enoch on 18 July 1956 and after stops at Saltcoats and Stevenston, kept the 27½ minute schedule to Paisley just reaching 60mph after Glengarnock. Many similar, punctual, but unremarkable runs are in the Rail Performance Society archive.

A few had gone to former Caledonian sheds – Stirling, Polmadie and Kingmoor. Some went to Carstairs in 1946 and five were placed at Inverness in 1947 (566, 592, 605, 619, 666) as replacements for the Highland 'Bens' and unsuperheated 'Dunalastairs'. However, this was not successful, and they were soon replaced by the later Pickersgill 4-4-0s. Then in 1948 603, 622 and 650 were moved to Kittybrewster taking over from ex-NBR D31s. No 40600, 40604, 40648 and 40663 followed in 1954 and 40617 and 40618 in 1955. They worked around Keith and the Maud to Peterhead branch and much local pick-up goods work.

In their final years many were stored, and records of unassisted runs are few. No 40632 of Nottingham was pressed into service on the 5.38pm St Pancras-Bedford commuter train in place of the rostered 'Black 5' on 1 December 1958, but made a very feeble effort, losing eight minutes unchecked to the first stop, Radlett, without exceeding 45mph and falling to 27 on the 1 in 176 to Elstree. Their main work of any note was assisting both Midland and West Coast route expresses when their loads exceeded the maximum for the rostered 4-6-0. I well remember standing on Willesden Junction station around 1955 and witnessing an up Manchester express tearing through at what I estimated to be well over 80mph with 40674 at the front swaying alarmingly as it took the curve towards Kensal Green Tunnel with a 'Jubilee' whose number I failed to catch. One of the pair of Bank Hall 2Ps, 40588 and 40684, regularly worked the 12.42pm Liverpool-Rochdale and the 3.45pm return until July 1960 when two Stanier 2-6-4Ts replaced them on these trains. No 40588 was withdrawn but 40684 then worked

No 40674 of Longsight leaving Manchester London Road with a stopping train for Stafford, 2 June 1954. (B.K.B. Green/MLS Collection)

No 40693 passing its home depot, Longsight, with the 1.18pm Manchester London Road-Crewe, 29 March 1955. (R. Gee/MLS Collection)

Carlisle's 40613 acting as station pilot at Carlisle Citadel station, as V2 60940 arrives from Edinburgh via the Waverley line with a train for Leeds while Stanier pacific 46240 *City of Coventry* stands on the centre road ready to change engines with the Polmadie Duchess that will bring the southbound *Royal Scot* into the platform, c1956.

Hurlford's 40570 leaves Kilmarnock with a stopping train for Ayr in June 1950. It is in that depot's typical excellent finish, recently repainted in the BR mixed traffic livery, but garnished with silver painted smokebox door hinges and straps and Caledonian blue numberplate. (J. Davenport/MLS Collection)

Ardrossan's 40609 at Glasgow St Enoch waiting to depart with an express for Largs, 8 June 1953. (A.C. Gilbert/MLS Collection)

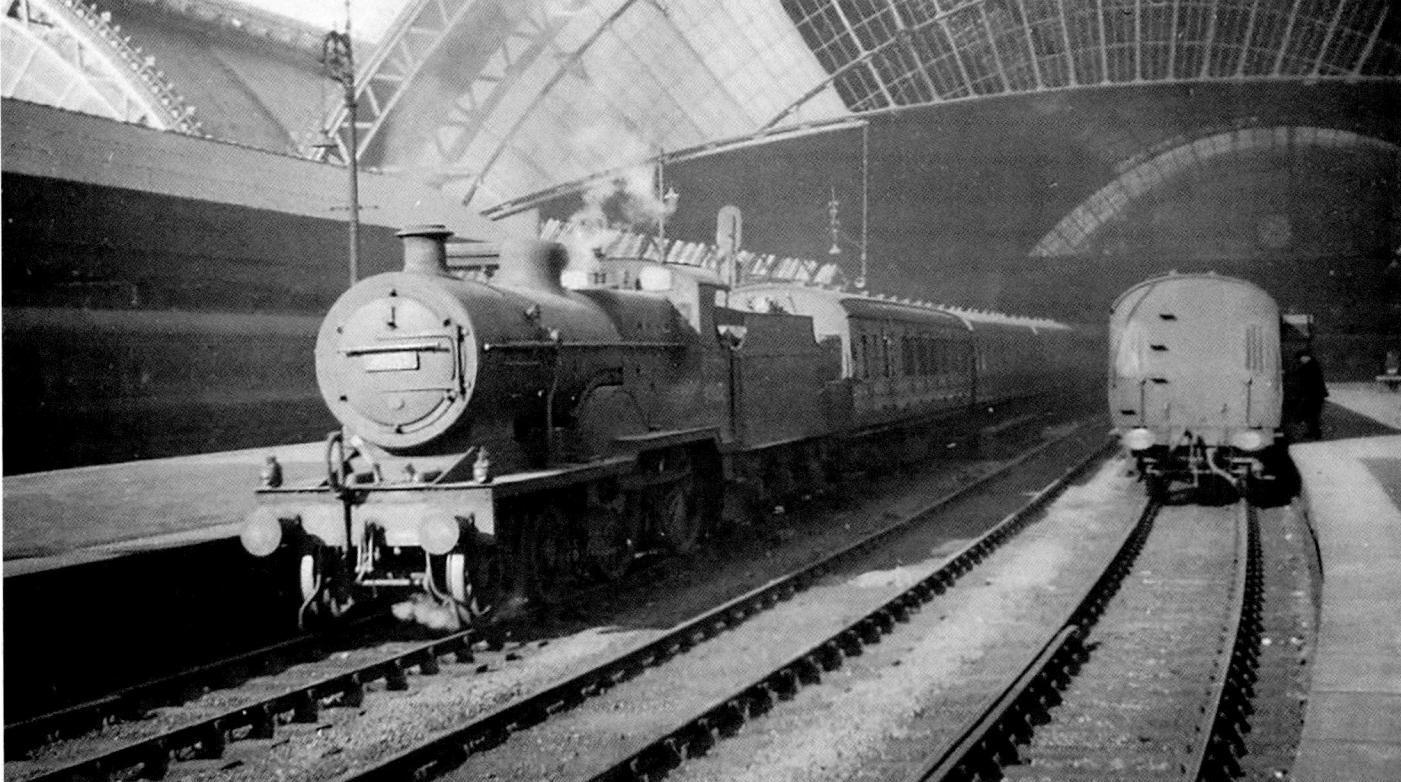

No 40667, another Ardrossan 2P, at Glasgow St Enoch with a train for Ardrossan, 20 September 1957. (David Maidment)

Ayr's 40574 at Kilmarnock with the 1.44pm stopping train to Troon, Prestwick and Ayr, 16 July 1958. Note AYR stencilled on the bufferbeam.
(David Maidment)

Stranraer's 40616 and 'Black 5' 45169 of Dumfries double-head a Stranraer-Dumfries-Carlisle train at New Galloway, 3 April 1958. Note STRANRAER stencilled on the bufferbeam. (MLS Collection)

the 9.13am Liverpool Exchange-Blackburn and 5.32pm return and throughout May 1961 was power for the 8.40am Rochdale-Liverpool, then the two-coach portion of a northbound Liverpool express to Preston where the Manchester portion took over.

I have a number of logs on the Midland main line, which I will quote to demonstrate that the crews were prepared to allow 2Ps to run fast (although whether they were being pushed by the train engine is problematical). No 40682 piloted 'Jubilee' 45561 *Saskatchewan* on a late running southbound train of 400 tons gross and the pair recovered time by running from Leicester to St Pancras ((99.1 miles) in 94 minutes 8 seconds (91 net) with some spectacular high speeds that must have caused some anxiety on the footplate of the 2P. The highlights were 85mph before Market Harborough, 86½ at Oakley on the descent from Sharnbrook and 82 at Hendon. No 40582 of Kentish Town and 40585 of Leicester seemed to feature prominently in such work in 1957. The latter engine piloted 45608 *Gibraltar* on the 6.33pm St Pancras-Derby on 31 October 1957 and the pair got their 335 ton train to Kettering (72 miles) in 74 minutes 43 seconds (69 net) with 82mph at Flitwick and a minimum of 53 at Sharnbrook summit. No 40582 with 45569 *Tasmania* on the up *Waverley* left Nottingham on time with 9 coaches, 315 tons, and passed Manton over five minutes late after a severe signal check. The pair then accelerated hard up to 88mph at Harringworth (the fastest I've seen recorded with a 2P), and after a 20mph check at Wellingborough, roared up Sharnbrook at 60mph and was then content with 75 at Bedford and 78 at Hendon. Signal checks ruined the run into the terminus, but net time was 118 minutes for the 123 miles. Nos 40582 and 45613 *Kenya* with a more substantial 420 ton load got from St Pancras to Kettering in 75 minutes with 80 at Flitwick, 43 at Sharnbrook and 75 at Wellingborough. Finally 40585 with 45696 *Arethusa* and a 10-coach 380 ton gross load left Kettering eight minutes late on a tight 70 minute schedule to St Pancras. The pair stormed Sharnbrook at 54mph, touched 81 before easing for Oakley troughs, sustained 60 on the long climb to Leagrave and then seemed to fall apart with speed down the temping 1 in 200s and 1 in 176 rising to no more than the mid-60s, both engines drifting (and both short of steam after earlier efforts?). The unchecked Kettering-St Pancras time was 71 minutes and arrival nine minutes late. By 1960, however, a number of redundant class 7 engines from the West Coast line, both 'Royal Scots' and 'Britannias', were transferred to the Midland and the need for double-heading reduced significantly. However, a Hurlford 2P regularly piloted the up *Thames-Clyde Express* from St Enoch to Dumfries up to the end of 1960, even after double-chimney 'A3s' took over from 'Royal Scots' and 'Britannias'.

No 40662 was withdrawn from Hurlford in 1954, but no more were withdrawn until 1957 when 40676 was withdrawn from Patricroft. No condemnations occurred in 1958 but the onslaught commenced in 1959, when forty-three were taken out of traffic, many of which had been stored for months at Uttoxeter and Derby on the Midland and at Hurlford in Scotland. At the end of 1960 the new Glasgow 'Blue Electrics' were withdrawn for six months following transformer failures and the Kilmarnock/Hurlford 2Ps had a brief reprieve when the Gresley V1 and V3 tanks that had replaced them were sent hurriedly back to Glasgow until some 2-6-4Ts were rustled up to make the 4-4-0s redundant again. Six 2Ps remained in action, four at Hurlford and two at Ardrossan until July 1961, but fourteen lay dead awaiting scrapping at Hurlford at that time. Three were suddenly based at the end of their lives at Watford, 40657 and 40683 moving there from Preston at the beginning of 1961 to join 40672 on engineering and local goods trains. The bulk, sixty-six, were withdrawn in 1961, leaving just fifteen which at the beginning of 1962 were allocated to:

Templecombe:	40563, 40564, 40634
Bath:	40696, 40697, 40700
Bescot:	40646, 40694
Patricroft:	40681
Watford:	40657, 40672
Ayr:	40638
Hurlford:	40664, 40665
Ardrossan:	40670

From this, it will be noted that the main activity at the end was on the Somerset and Dorset, an activity they shared with the remaining Midland 2P, 40537. Their last active season was the summer of 1961 when they assisted many of the heavy holiday expresses from the

West Midlands to Bournemouth - they were replaced in the following year by BR Standard 4 4-6-0s for those duties. No 40563 was withdrawn in May the following year, and was the longest lived of the LMS 2Ps, thirty-four years. Watford's 40672 was employed almost exclusively on saloon specials for the Watford based Divisional Civil Engineer, although it is known to have substituted for the usual 2-6-4T on a Watford-Euston commuter train on a few occasions. The last to be withdrawn was 40670. All were gone by the end of 1962 and none was preserved.

Bank Hall's 40588, withdrawn in 1961, at Castleton, c1960.
(R.S. Greenwood/ MLS Collection)

No 40588 again, shortly before withdrawal in 1961, at Bolton.
(MLS Collection)

Bank Hall's 40684 assists 'Rebuilt Patriot' 45534 *E. Tootal Broadhurst* with a Trans-Pennine express at Orrell West, 25 March 1961. The 2P was withdrawn later that year. (A.C. Gilbert/MLS Collection)

The last active LMS 2Ps were on the Somerset & Dorset. Here 40564 is assisting S&D 7F 2-8-0 53810 with a heavy Summer Saturday holiday train approaching Masbury summit, 2 September 1961. (MLS Collection)

Chapter 8
CONCLUSIONS

R.E. Charlewood, in his 1907 *Railway Magazine* article 'British Locomotive Practice and Performance', was somewhat sceptical about the Midland Railway practice of frequent lightly loaded expresses which seemed to require double-heading as traffic increased in the first decade of the twentieth century. In earlier articles, he and various correspondents had been debating the merits of the different railway 'Singles' versus the newer four-coupled engines. Johnson was still building 'Singles' years after he had introduced the first 4-4-0s and they were apparently coping easily with the lightly loaded Midland expresses although their low adhesion were disadvantages on starting and the heaviest gradients where the engine could not get a run at it. He seemed surprised that the Johnson 4-4-0s had only 1,193sq ft of heating surface compared with 2,000-2,500sq ft of many of the contemporary express engines of other companies and therefore seemed little more effective than the 'Singles'. Although Midland expresses were as fast as any of the competing railways, the services tended to be frequent and light in comparison, especially with the LNWR and GNR.

The small-boilered 4-4-0s of the 1880s were limited to 160 tons on the fastest expresses but by 1907 many had been rebuilt with 175lbs psi 'H' boilers. He asked his readers:

'Has the abandonment of the small standard boiler in former use on the Midland Railway resulted in any reduction of piloting on that railway? Most undoubtedly it has. But old habits are not easily got rid of and although there is an enormous reduction in the aggregate amount of MR piloting, the practice is still resorted to at times when, in my opinion, it might and should be avoided.'

He then gave an example of the evening Scottish-London express allowed 131 minutes for the non-stop 123¾ miles from Nottingham, 56.7mph average booked for three 12-wheel and four 6-wheel vehicles and one of the new 'Belpaires' (G7 boiler). In his example, it was piloted by a Johnson 7ft 4in 'Single' and the pair completed the journey including a special Luton stop in 128 minutes (124 net). He did not doubt that the train engine could have done the run unassisted, but clearly his confidence was not shared by the Nottingham locomotive authorities.

Charlewood's predecessor, Charles Rous-Marten, had also been a fan of the 'Singles', and in the early years of the twentieth century, took great interest in the potential of compounding, especially the Churchward trials of a De Glehn Nord compound on the Great Western. The Smith system developed on the North Eastern was also admired by him and he was very impressed with the prototype Midland five, 2631-2635, which used this system. However, Deeley challenged his comments saying 'the Smith reducing valve arrangements frequently failed, so I fitted the engines with a much more simple starting arrangement. There are no "Smith" compounds on the Midland Railway. They have all been altered…'

Despite – or because of – this, Rous-Marten was very complimentary about those first five compounds, saying:

'The work done was, without any exception or qualification, the best I ever saw on the Midland Railway and this not only on the London-Leicester section, where an express of twenty-one coaches weighing 370 tons was hauled to time without pilot aid, but also on the Leeds-Carlisle section, where the Compound made quite light of the long banks of 1 in 100 and steeper for fifteen miles continuously, hauling unassisted 240 to 250 tons behind the tender.'

Cecil J. Allen also commented on the fact that the Midland Railway was still using 'Singles' in 1910, providing, in his words, 'highly efficient assistance'. I'm not sure how efficient it was to use pilot engines so frequently. The 'Singles' were light on coal but required the extra crew. They were apparently still rostered to the lighter trains single-handed, CJA quoting the 72 mile St Pancras-Kettering section covered in 84 minutes with 180 tons and No 615 completing the 99-mile Leicester-London route in 107 minutes with 150 tons. They were also used as standbys, with one rescuing an ailing 769 (3P) at Bedford on a down 210-ton Sheffield train and assisting the poor-steaming engine to recover twelve of the lost fifteen minutes between Leicester and Sheffield. CJA commented, however, very favourably on the performance of the Compounds compared with the Ivatt Atlantics:

> 'The GNR engines are slow to start and uphill. The Midland are working their 3-cylinder engines at high pressure on starting and are able to accelerate rapidly and also, with a judicious proportion of high pressure steam regulated by an automatic valve, are able to tackle the much heavier grades of the line at greatly increased speeds. In this way, though loads are considerably lighter *(than the GNR)*, the faster schedules of the Midland are daily observed with almost unnerving punctuality – and without excessive downhill speeds.'

In the period before the First World War, the Midland was running 11,000 miles daily at 40mph+, but with indifferent punctuality (despite C.J. Allen's comments above) and much double-heading. They had 5,800 (115 runs) at over 50mph. The fastest schedules were from St Pancras to Kettering, 76 minutes for 72 miles (56.8mph), the longest non-stop runs, 196 miles St Pancras to Leeds. Speeds were as high as any other railway, 80+mph being common, 90 rare. Services on the Midland Railway trunk route were frequent. In 1910, there were twenty-eight daily trains from St Pancras as far as Trent and twelve to Manchester. The marketing concept seems very similar to the Virgin Voyager cross-country services after privatisation through Birmingham New Street, the Midland policy largely dictated by the operators stemmed from the 1909 Paget Control system and strict limitation of loads. Subsequent motive power performance never met the pre-1909 standards, as the new load restrictions or double-heading meant that the Midland 4-4-0s were rarely called upon to produce the power output that the LNWR, GNR or GWR found necessary to run their fast but heavier unassisted trains. As a result of the Midland engines being driven routinely well within their capabilities, they were economical in terms of water and coal consumption. The 4-4-0s, especially the 4P Compounds, were found by train timers to be fast uphill, but easy, with light steam, downhill.

Whether the reduced running and maintenance costs of the Midland 4-4-0s offset the increased staffing costs of more locomotives and double-heading is uncertain. Some costs quoted before the Paget 1909 timetable 'revolution' were (1906): 11.1d per loaded engine mile for Midland & GWR, 10.6 for GNR, 12.5 LNWR, 14.2 NER. The number of locomotives the Midland needed to convey similar loads compared with the other main companies reflects their small engine policy.

Number of locomotives per track mile:
Midland	0.83
Great Western	0.46
North Eastern	0.4
LNWR	0.57

Loaded train miles per loco per day:
Midland	28
Great Western	36
North Eastern	26
Great Northern	28
LNWR	29

This suggests that the Great Western had the most efficient motive power policy, but this does not take into account the passenger loadings on the trains and whether the light, fast frequent service of the Midland was the most profitable commercial policy. This is obscured by the fact that the Midland and Great Western railways in particular made most of their profit from extensive coal haulage from the South Wales and South Yorkshire and Nottinghamshire coal fields. I think the Midland policy is defensible up to the First World War.

However, the perpetuation of the policy after the war, and especially after the Grouping, seems weak and lacking in ambition compared with the other three newly formed companies. In January 1923, the most powerful locomotives the Midland passed to the LMS were just 45 Compounds out of a fleet of 3,000 engines. Compare this with the GW's 73 'Stars' and 70 + 'Saints', the Southern's Urie N15s and all Drummond's 4-6-0s (admittedly not the most efficient of engines, but powerful), the GN and NE's progression from Atlantics to Pacifics, and the 4-6-0s of the LNWR and L&Y. By 1926, the GW had added 40 Castles, the SR 40 Maunsell N15s and the LNER a good number of A1 pacifics. The LMS just built 190 extra 4P Compounds.

The LMS management was very cost-conscious and around 1924 introduced costing of individual locomotives, both coal and oil consumption in traffic and maintenance and repair costs. The early Johnson 4-4-0s were economical and fast but produced little more power or speed than the 'Singles' and although improved by boiler rebuilding – the Deeley H in the 1904-8 period and superheating from 1912 – never achieved their potential had they been rebuilt with long travel and long lap valve gear as the Churchward engines and Maunsell's similar sized rebuilds of Wainwright's Ds and Es. The 3P 'Belpaires' were fine in their time but had been superseded by 1910. The Midland Compounds were excellent engines and did all required of them until around 1923, but the perpetuation of them for the West Coast services was in hindsight a mistake. They continued to perform economically on light services like the LMS London-Birmingham services as well as on the Midland for a while but were outclassed and displaced by Stanier's 4-6-0s and then relegated to secondary services where they were 'run of the mill' without regular crews and their reliability and their running and maintenance costs suffered. Of the LMS 2P 4-4-0s, the best that can be said for them is that they performed useful secondary work economically – in fact under the LMS engine costing system they recorded the most economical figures of any LMS passenger engines in the 1930s.

Whether their investment costs ever gave an adequate return by reducing the maintenance costs of older pre-Grouping engines they replaced must be questionable. An article entitled 'Living with L.M. Locomotives' in *Trains Illustrated* in November 1957 by an author who preferred to remain anonymous but had been an apprentice at Derby Works in the 1930s, was damning on the weaknesses of the Midland and LMS Fowler engines. Some quotes: 'Every Midland engine was (and is) just about as hot-box prone as ingenuity can make it.' And: 'Midland cylinder and front-end design can only be described as appalling (with the sole exception of the '990' 4-4-0s).... Small, short travel piston valves, allied to the most tortuous ports and passages that the foundry could cast, ensured that steam had as difficult a job as human ingenuity could devise to get in and out of the cylinders. Hence the "Derby roar", otherwise known as "much ado about nothing".' Clearly the writer had developed from personal hard experience a strong antipathy to these engines but the fact remains that on the road they did the job that the Midland and LMS authorities required of them, and in their prime, at an acceptable cost.

Appendix
DIMENSIONS, WEIGHT DIAGRAMS & STATISTICS

Midland Class 2
1312 series
Dimensions

Cylinders (2 inside)	17½in x 26in
Coupled wheel diameter	6ft 6in
Bogie wheel diameter	3ft 3in
Boiler pressure	140lb psi
Heating surface	1,215sq ft
Grate area	17.5sq ft
Axleweight	14 tons
Weight: Engine	41 tons 19 cwt
Tender	35 tons 3 cwt
Total	77 tons 2 cwt
Water capacity	2,800 gallons
Coal capacity	3½ tons
Tractive effort	10,800lbs

1327 series
Dimensions

As above apart from:

Cylinders (2 inside)	18in x 26in
Coupled wheel diameter	7ft 0in
Bogie wheel diameter	3ft 6in
Heating surface	1,313sq ft (later changed to 1,260sq ft)
Water capacity	2,960 gallons

Weight diagrams

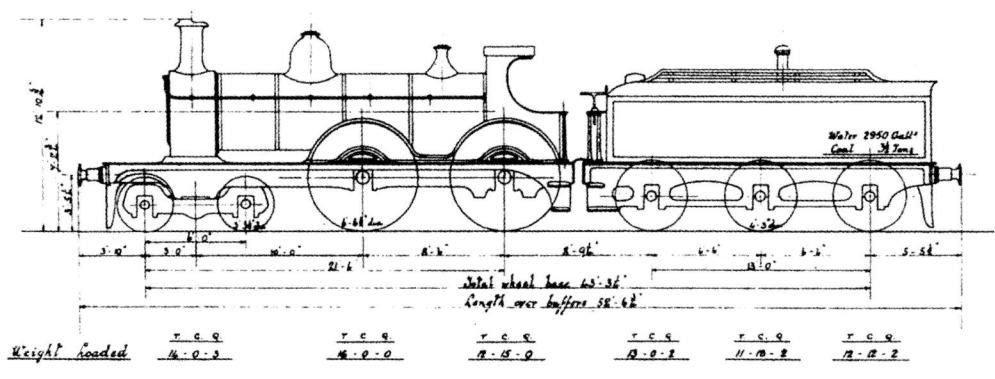

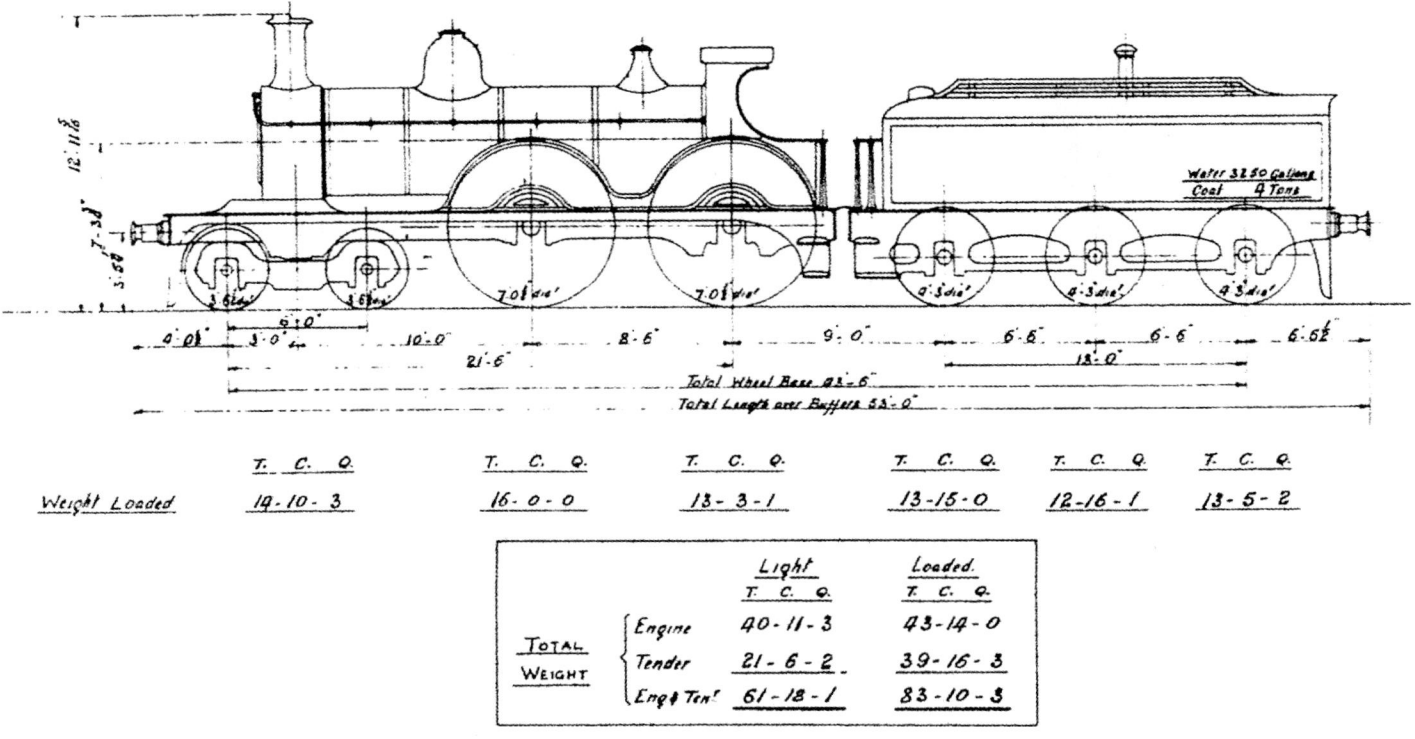

Statistics

No.	Built	1907	First depot	Last depot	Withdrawn	
1312	11/76	300	Manchester (Cornbrook)	Sheffield	12/24	
1313	11/76	301	Cornbrook	Toton	8/20	
1314	12/76	302	Cornbrook	Sheffield	1/20	
1315	12/76	303	Cornbrook	Sheffield	1/24	
1316	12/76	304	Cornbrook	Sheffield	3/25	
1317	12/76	305	Cornbrook	Sheffield	9/26	
1318	1/77	306	Cornbrook	Derby	11/30	
1319	1/77	307	Cornbrook	Brunswick	5/11	
1320	2/77	308	Cornbrook	Sheffield	12/28	
1321	2/77	309	Cornbrook	Brunswick	11/12	
1327	2/77	310	Cornbrook	Skipton	3/28	New frames 1898
1328	6/77	311	Cornbrook	Peterborough East	11/34	New frames 5/11
1329	6/77	312	London	Skipton	9/20	
1330	6/77	313	London	Skipton	8/20	
1331	6/77	314	London	Skipton	12/28	New frames 6/16
1332	6/77		London	Cornbrook	11/04	

No.	Built	1907	First depot	Last depot	Withdrawn	
1333	6/77	315	London	Manchester	11/12	
1334	6/77	316	London	Manchester	11/12	
1335	7/77	317	London	Manchester	12/24	
1336	7/77		London	Saltley	3/04	
1337	8/77	319	London	Sheffield	12/24	
1338	8/77	320	London	Sheffield	9/28	
1339	8/77	321	Derby	Sheffield	12/18	
1340	8/77	322	Leeds	Sheffield	8/20	
1341	8/77	323	Leeds	Sheffield	12/28	New frames 3/09
1342	10/77	324	Leeds	Sheffield	11/12	
1343	10/77	325	Leeds	Leicester	7/28	
1344	10/77	326	Leeds	Sheffield	7/28	New frames 5/14
1345	10/77	327	Leeds	Sheffield	6/25	
1346	11/77	318	Brunswick	Sheffield	8/19	

1562 series

Dimensions

Cylinders (2 inside)	18in x 26in	Axleweight	15 tons 14 cwt
Coupled wheel diameter	6ft 9in	Weight: Engine	41 tons 19 cwt
Bogie wheel diameter	3ft 6in	Tender	35 tons 3 cwt
Stephenson valve gear		Total	77 tons 2 cwt
Boiler pressure	'B' type, 140lb psi	Water capacity	2,950 gallons
Heating surface	1,260sq ft	Coal capacity	3½ tons
Grate area	17.5sq ft	Tractive effort	14,560lbs

Weight diagram

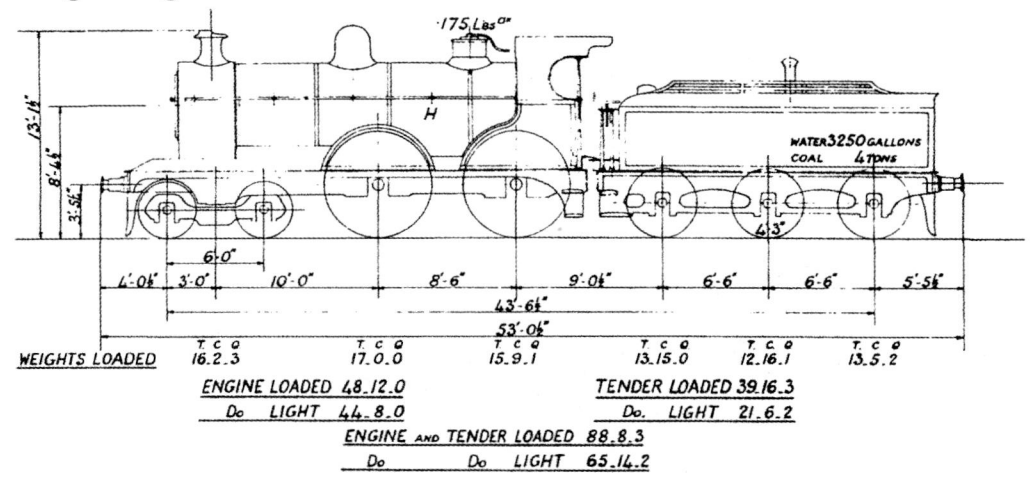

Statistics

No.	Built	1907	H boiler		G7 boiler	First depot	Last depot	Withdrawn
1562	9/82	328	4/07		9/09	Leicester	Derby	1/28
1563	9/82	329	3/06		5/11	Leicester	Buxton	5/30
1564	10/82	330	11/06		9/09	Leicester	Brunswick	11/32
1565	10/82	331	4/07			Leicester	Brunswick	3/28
1566	10/82	332	11/06		8/23*	Leicester	Bristol	12/59 (40332)
1567	10/82	333	12/06			Leicester	Brunswick	12/26
1568	11/82	334	2/07		9/10	Kentish Town	Brunswick	4/30
1569	11/82	335	1/07			Kentish Town	Brunswick	6/27
1570	11/82	336	12/07		9/09	Kentish Town	Brunswick	8/32
1571	11/82	337	12/06	6/09/3/23*		Kentish Town	Hasland	4/58 (40337)
1572	12/82	338	2/07		5/09	Carlisle	Manchester	2/37
1573	2/83	339	2/07		6/09	Carlisle	Manchester	6/28
1574	2/83	340	3/07		6/09	Carlisle	Buxton	8/30
1575	2/83	341	12/06			Carlisle	Manchester	12/26
1576	2/83	342	3/07		9/11	Carlisle	Manchester	12/33
1577	3/83	343	3/07			Carlisle	Manchester	1/26
1578	3/83	344	3/08			Carlisle	Manchester	12/26
1579	3/83	345	12/06			Carlisle	Manchester	8/25
1580	3/83	346	5/07		8/09	Carlisle	Manchester	2/28
1581	4/83	347	6/07		9/09	Carlisle	Manchester	12/32
1657	10/83	348	12/06			Manchester	Manchester	4/27
1658	10/83	349	1/07		11/10	Manchester	Manchester	10/27
1659	10/83	350	1/07		6/09	Manchester	Manchester	3/32
1660	10/83	351	12/06	5/10/4/23*		Manchester	Leeds	12/53 (40351)
1661	11/83	352	1/07			Manchester	Manchester	7/26
1662	10/83	353	1/08		5/23*	Manchester	Wellingborough	7/53 (40353)
1663	10/83	354	5/07			Manchester	Manchester	6/26
1664	11/83	355	5/07			Manchester	Manchester	9/26
1665	11/83	356	1/08	5/11/ 3/23*		Manchester	Carlisle (Upp)	4/57 (40356)
1666	11/83	357	3/07			Manchester	Manchester	1/26

* Rebuilt in 1923 with class 483 superheater boiler

1667 series
Dimensions
As for 1562 class apart from:
Cylinders	19in x 26in
Coupled wheels	7ft 0½in
Boiler pressure	140lbs psi at first, 160lbs psi from 1886

Joy's valve gear & motion

Statistics

No.	Built	First depot	Last depot	Withdrawn (replaced by new engines with same no.)
1667	5/84	*	Manchester	3/97
1668	6/84	*	Manchester	12/96
1669	6/84	*	Manchester	9/98
1670	6/84	*	Manchester	4/01
1671	7/84	*	Manchester	10/98
1672	8/84	*		10/96
1673	8/84	*		5/01
1674	8/84	*		6/01
1675	9/84	*	Kentish Town	10/96
1676	10/84	*	Kentish Town	3/97

* Kentish Town (5), Nottingham (3) & Derby (2), numbers not specified.

1738 series
Dimensions
Cylinders (2 inside)	18in x 26in
Coupled wheel diameter	7ft 0½ in
Bogie wheel diameter	3ft 6in
Stephenson valve gear	
Boiler pressure	'B' type, 160lb psi
Heating surface	1,260sq ft
Grate area	17.5sq ft
Axleweight	15 tons 14 cwt
Weight: Engine	41 tons 19 cwt
Tender	35 tons 3 cwt
Total	77 tons 2 cwt
Water capacity	2,950 gallons
Coal capacity	3½ tons
Tractive effort	13,957lbs

Dimensions, Weight Diagrams & Statistics • 307

Weight diagram

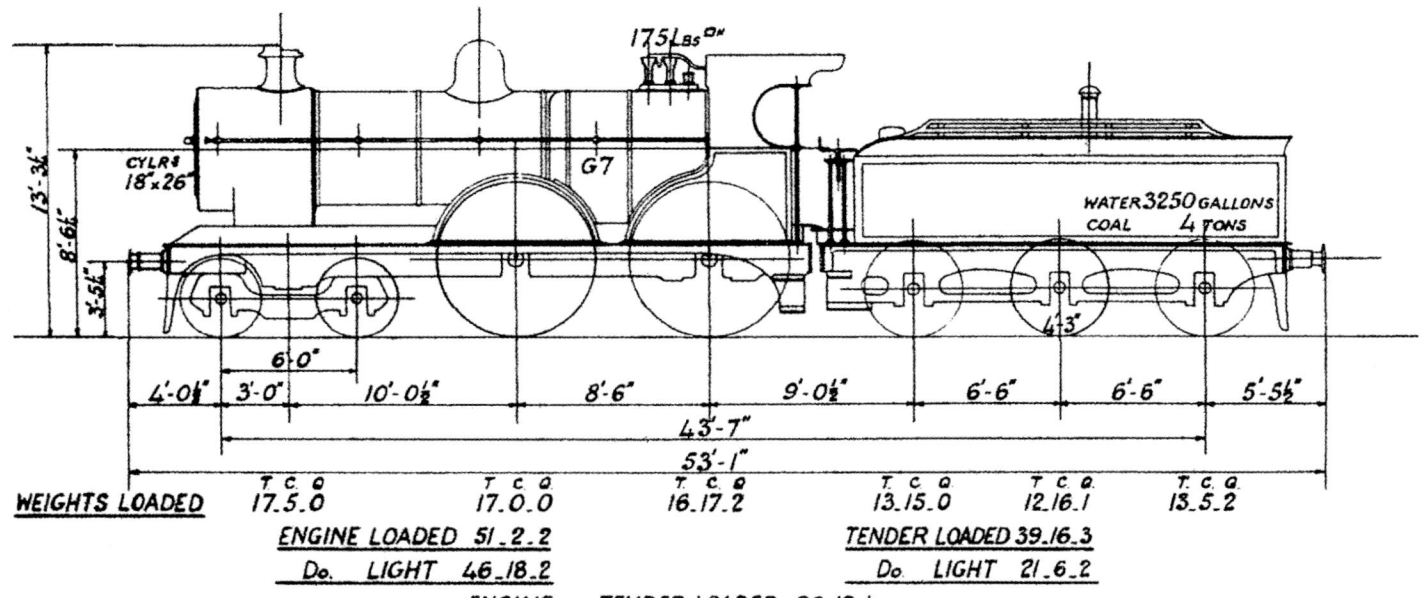

Statistics

No.	Built	1907	H boiler	G7 boiler	First depot	Last depot	Withdrawn
1738	11/85	358	10/07	7/09	Kentish Town	Belle Vue	7/29
1739	11/85	359	11/06	4/23*	Kentish Town	Hasland	2/54 (40359)
1740	12/85	360	3/07		Kentish Town	Leeds	12/26
1741	12/85	361	5/07	10/10	Kentish Town	Leeds	12/29
1742	12/85	362	2/07	11/23*	Kentish Town	Lancaster	12/56 (40362)
1743	1/86	363	2/07	4/11	Kentish Town	Leeds	9/28
1744	1/86	364	12/07	1/24*	Kentish Town	Burton	6/56 (40364)
1745	1/86	365	2/07	10/10	Bedford	Leeds	12/36
1746	1/86	366	12/06	3/11	Kentish Town	Leeds	8/31
1747	2/86	367	1/07		Kentish Town	Leeds	6/26
1748	6/86	368	1/07	3/11	Carlisle	Saltley	12/39
1749	6/86	369	5/07	11/10	Carlisle	Bourneville	12/40

No.	Built	1907	H boiler	G7 boiler	First depot	Last depot	Withdrawn
1750	6/86	370	2/07	11/22*	Leeds	Toton	9/50 (40370)
1751	8/86	371	12/06		Leeds	Worcester	6/27
1752	8/86	372	9/06	6/11	Leeds	Hasland	7/40
1753	8/86	373	11/06		Leeds	Sheffield	11/25
1754	9/86	374	10/06	10/10	Leeds	Burton	5/35
1755	9/86	375	1/07		Leeds	Saltley	7/26
1756	9/86	376	2/07	8/11	Leeds	Burton	11/35
1757	12/86	377	1/07	5/23*	Kentish Town	Llandudno Jn	9/55 (40377)

* rebuilt with superheated boiler to class 483

The 1808 series

Dimensions

Cylinders (2 inside)	18in x 26in
Coupled wheel diameter	6ft 6in
Bogie wheel diameter	3ft 3in
Stephenson valve gear	
Boiler pressure	'B' type, 160lb psi
Heating surface	1,260sq ft
Grate area	17.5sq ft
Axleweight	15 tons 14 cwt
Weight: Engine	41 tons 19 cwt (with H boiler (46 tons 16 cwt))
Tender	35 tons 3 cwt
Total	77 tons 2 cwt
Water capacity	2,950 gallons
Coal capacity	3½ tons
Tractive effort	15,019lbs

Statistics

No.	Built	1907	H boiler	G7 boiler	First depot	Last depot	Withdrawn	
1808	4/88	378	10/05	11/10	Belle Vue	Derby	10/47	
1809	4/88	379	12/04	4/11	Belle Vue	Sheffield	12/37	
1810	5/88	380	11/05	2/11	Belle Vue	Sheffield	12/32	
1811	5/88	381	11/04	8/10	Belle Vue	Sheffield	3/36	
1812	5/88	382	6/04	6/11	Belle Vue	Sheffield	11/36	
1813	6/88	383	6/04	12/09	Brunswick	Derby	7/52	(40383)
1814	6/88	384	6/04	5/10	Brunswick	Sheffield	10/25	

No.	Built	1907	H boiler	G7 boiler	First depot	Last depot	Withdrawn	
1815	6/88	385	12/04	5/10	Brunswick	Saltley	9/49	
1816	6/88	386	12/04	5/09	Brunswick	Sheffield	10/31	
1817	6/88	387	10/04	6/11	Brunswick	Sheffield	12/36	
1818	7/88	388	3/05	9/09	Brunswick	Sheffield	3/30	
1819	7/88	389	5/05	9/10	Carnforth	Sheffield	2/35	
1820	8/88	390	9/05	5/10	Carnforth	Sheffield	12/32	
1821	9/88	391	4/05	2/10	Carnforth	Manningham	9/49	
1822	9/88	392	6/04	1/12	Carnforth	Sheffield	12/32	
80	6/91	393	5/04	11/09	Carnforth	Sheffield	12/28	
81	6/91	394	5/05	11/23*	Carnforth	Sheffield	3/49	
82	6/91	395	12/04	6/22*	Carnforth	Nottingham	10/54	(40395)
83	6/91	396	6/05	5/23*	Carnforth	Burton	2/61	(40396)
84	6/91	397	12/04	10/22*	Carlisle	Brunswick	3/51	(40397)
85	6/91	398	6/04		Carlisle	Saltley	9/26	
86	6/91	399	6/04		Carlisle	York	10/25	
87	6/91	400	6/04	10/22*	Carlisle	York	1/49	
11	8/91	401	2/05	7/22*	Nottingham	Walsall	9/53	(40401)
14	8/91	402	10/04	6/22*	Nottingham	Leicester	10/60	(40402)

* rebuilt with superheater boiler (483 class)

The 2183 series
Dimensions

Cylinders (2 inside)	18½in x 26in
Coupled wheel diameter	7ft 0in
Bogie wheel diameter	3ft 6in
Stephenson valve gear	
Boiler pressure	'D' type, 160lb psi
Heating surface	1,205sq ft
Grate area	17.5sq ft
Axleweight	16 tons 19 cwt
Weight: Engine	47 tons 4 cwt
Tender	37 tons
Total	84 tons 4cwt
Water capacity	3,250 gallons
Coal capacity	4 tons
Tractive effort	15,019lbs

Weight diagram

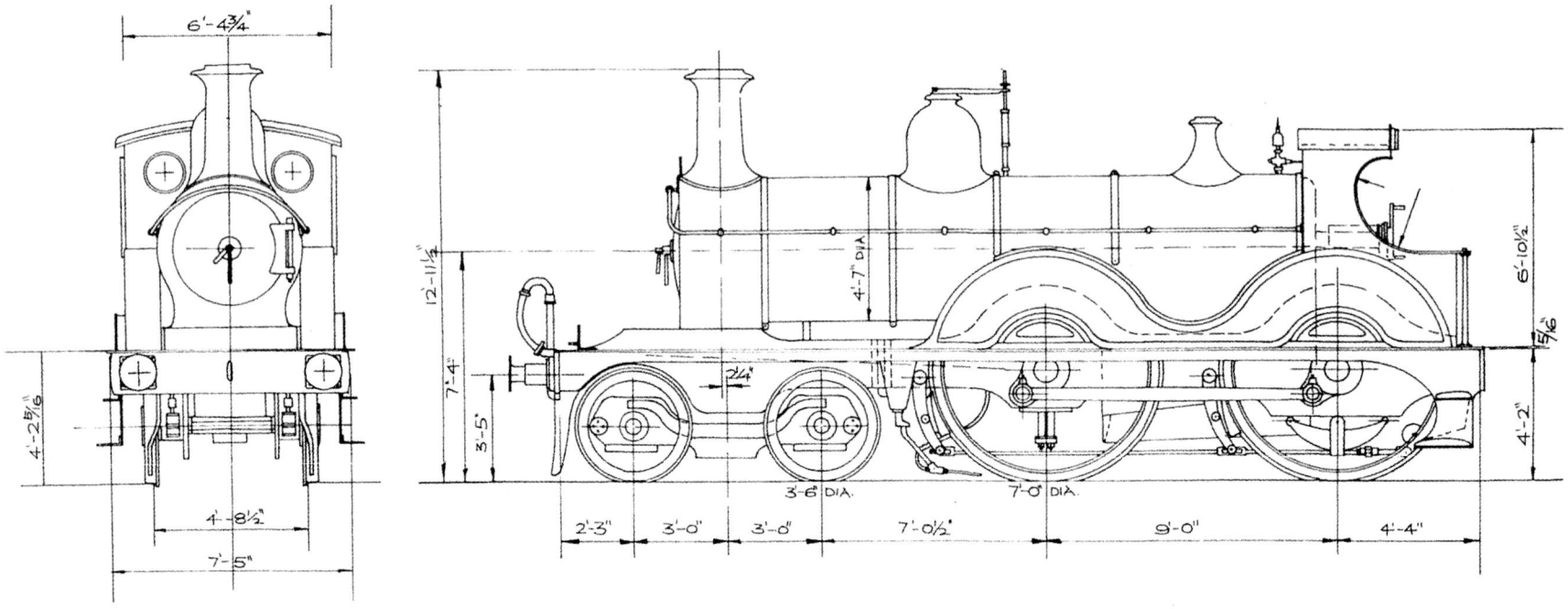

Statistics

No.	Built	1907	H boiler	483 boiler	First depot	Last depot	Withdrawn	
2183	4/92	403	12/7	12/20	Nottingham		6/50	
2184	4/92	404	2/08	2/18	Nottingham	Derby	7/57	(40404)
2185	4/92	405	2/06	6/14	Nottingham	Heaton Mersey	3/55	(40405)
2186	4/92	406	5/06	6/14	Nottingham	Normanton	8/52	(40406)
2187	5/92	407	7/06	7/14	Nottingham	Derby	6/58	(40407)
2188	5/92	408	5/06	5/14	Nottingham		12/48	
2189	5/92	409	6/06	9/19	Nottingham	Skipton	7/57	(40409)
2190	5/92	410	5/06	1/21	Nottingham	Preston	2/53	(40410)
2191	5/92	411	6/07	9/18	Nottingham	Nottingham	2/61	(40411)
2192	5/92	412	5/06	5/14	Nottingham	Derby	5/59	(40412)
2193	5/92	413	6/07	9/18	Kentish Town	Kentish Town	8/58	(40413)
2194	5/92	414	4/06	7/14	Kentish Town	Burton	1/57	(40414)
2195	5/92	415	5/06	9/14	Kentish Town	Nottingham	6/52	(40415)
2196	5/92	416	2/06	8/14	Kentish Town	Derby	5/59	(40416)
2197	5/92	417	6/06	12/17	Kentish Town	Nottingham	2/52	(40417)
2198	7/92	418	7/06	6/14	Bedford	Derby	1/57	(40418)

No.	Built	1907	H boiler	483 boiler	First depot	Last depot	Withdrawn	
2199	7/92	419	4/06	9/19	Bedford	Crewe	9/55	(40419)
2200	7/92	420	2/06	7/16	Bedford	Kentish Town	10/58	(40420)
2201	7/92	421	5/06	2/18	Bedford	Nottingham	1/61	(40421)
2202	7/92	422	1/06	4/22	Bedford	Skipton	10/53	(40422)
156	8/96	423	2/06	10/18	Nottingham	Gloucester	1/53	(40423)
157	8/96	424	3/06	10/14	Nottingham	Mansfield	4/51	(40424)
158	8/96	425	4/07	5/14	Nottingham	Crewe	12/53	(40425)
159	9/96	426	7/06	6/16	Nottingham	Bristol	11/57	(40426)
160	9/96	427	8/06	3/15	Nottingham		5/50	(40427)

The 2203 series

Dimensions

Cylinders (2 inside)	18½in x 26in
Coupled wheel diameter	6ft 6in
Bogie wheel diameter	3ft 3in
Stephenson valve gear	
Boiler pressure	'D' type, 160 lb psi
Heating surface	1,205sq ft
Grate area	19.6sq ft
Axleweight	16 tons 19 cwt
Weight: Engine	47 tons 4 cwt
Tender	37 tons
Total	84 tons 4 cwt
Water capacity	3,250 gallons
Coal capacity	4 tons
Tractive effort	15,019lbs (460 & 465 G7 boiler, 15,879lbs)

Weight diagram

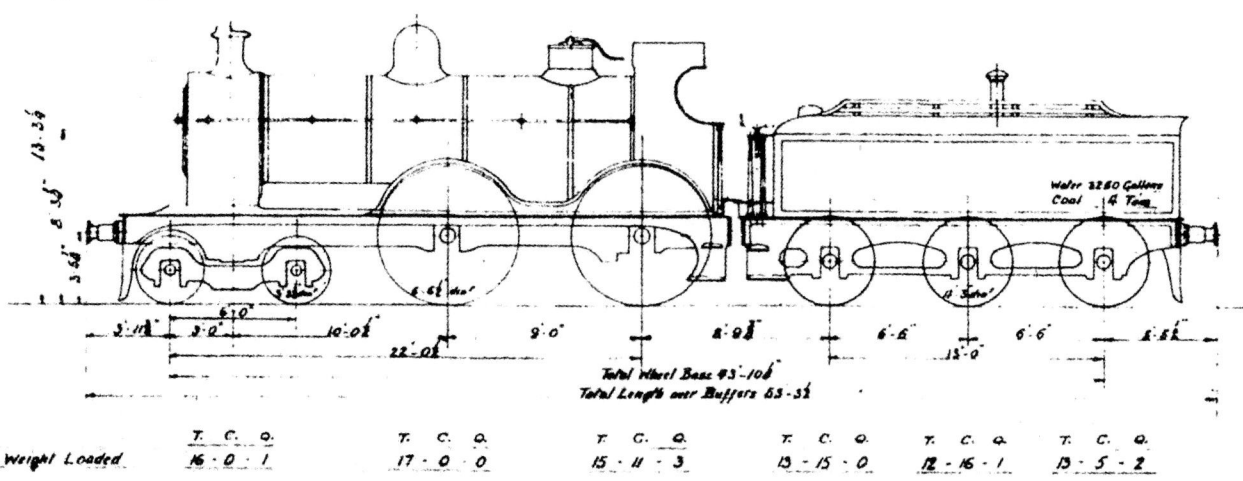

Statistics

No.	Built	1907	H boiler	483 boiler	First depot	Last depot	Withdrawn	
2203	1/93	428	5/05		Leeds	Nottingham	4/27	
2204	2/93	429	5/05		Leeds	Saltley	6/26	
2205	2/93	430	6/04	1/21	Leeds	Heaton Mersey	4/52	(40430)
2206	2/93	431	6/05		Leeds	Saltley	8/25	
2207	2/93	432	3/05	10/16	Belle Vue	Burton	10/53	(40432)
2208	2/93	433	11/04	6/15	Belle Vue	Kentish Town	11/57	(40433)
2209	2/93	434	4/06	7/16	Belle Vue	Patricroft	11/56	(40434)
2210	2/93	435	3/06		Belle Vue	Saltley	9/27	
2211	3/93	436	11/05	9/14	Belle Vue	Burton	9/54	(40436)
2212	3/93	437	3/06	12/14	Belle Vue		5/49	
2213	3/93	438	2/06	7/23	Belle Vue	Patricroft	8/54	(40438)
2214	3/93	439	8/05	9/23	Belle Vue	Toton	1/61	(40439)
2215	4/93	440	7/05		Belle Vue	Leeds	6/26	
2216	4/93	441	1/06		Belle Vue	Leeds	10/25	
2217	4/93	442	3/05		Belle Vue	Carlisle	10/26	
184	2/94	443	11/04	1/15	Carlisle	Toton	1/61	(40443)
185	2/94	444	3/05	10/14	Carlisle	Leeds	7/53	(40444)
186	2/94	445	2/05		Carlisle	Carlisle	6/26	
187	3/94	446	12/05	3/21	Carlisle		6/50	
188	3/94	447	3/06	3/20	Carlisle	Nottingham	5/58	(40447)
189	4/94	448	3/05	10/19	Carlisle	Upperby	11/55	(40448)
190	4/94	449	6/05		Carlisle	Lancaster	3/25	
191	4/94	450	5/06	11/23	Carlisle	Patricroft	3/57	(40450)
192	4/94	451	5/06		Carlisle	Lancaster	2/26	
193	5/94	452	11/04	3/20	Carlisle	Leicester	1/61	(40452)
194	6/94	453	3/06	2/17	Carlisle	Burton	10/62	(40453)
195	6/94	454	2/05	5/22	Carlisle	Nottingham	9/60	(40454)
196	6/94	455	6/05	12/16	Carlisle	Manningham	7/54	(40455)
197	7/94	456	9/04	3/20	Carlisle		11/49	
198	7/94	457	5/06		Carlisle	Nottingham	6/26	
199	8/94	458	3/05	2/15	Carlisle	Nottingham	2/57	(40458)
161	8/94	459	9/05	9/22	Leicester		12/49	
162	9/94	460	4/06	10/10 (G7)	Leicester	Skipton	2/30	
163	9/94	461	2/06	6/15	Leicester	Nottingham	2/59	(40461)
164	9/94	462	6/05	10/16	Leicester	Walsall	4/51	(40462)
230	8/95	463	3/05	3/15	Leeds	Bourneville	7/56	(40463)

No.	Built	1907	H boiler	483 boiler	First depot	Last depot	Withdrawn	
231	9/95	464	3/05	12/14	Leeds	Leicester	3/58	(40464)
232	9/95	465	11/04	5/10 (G7)	Leeds	Skipton	11/31	
233	10/95	466	1/06	12/14	Leeds		3/49	
234	10/95	467	5/05		Leeds	Skipton	6/26	
235	11/95	468	10/04	8/16	Leeds		5/50	
236	11/95	469	1/06		Carnforth	Skipton	12/26	
237	11/95	470	4/05	4/17	Carnforth	Hellifield	5/51	(40470)
238	12/95	471	12/04	10/22	Carnforth	Stafford	1/52	(40471)
239	12/95	472	3/06	11/23	Carnforth	Skipton	9/55	(40472)

The 2581 series

Dimensions

Cylinders (2 inside)	18½in x 26in
Coupled wheel diameter	6ft 6in
Bogie wheel diameter	3ft 3in
Stephenson valve gear	
Boiler pressure	'B' type, 160lb psi
Heating surface	1,260sq ft
Grate area	17.5sq ft
Axleweight	15 tons 14 cwt
Weight: Engine	41 tons 19 cwt (with H boiler (46 tons 16 cwt))
Tender	37 tons
Total	78 tons 19 cwt
Water capacity	3,230 gallons
Coal capacity	4 tons

Statistics

No.	Built 1907		H boiler	483 class*	First depot	Last depot	Withdrawn
2581	3/00	473	3/05		Derby	Leeds	6/27
2582	3/00	474	8/05		Derby	Leeds	10/25
2583	3/00	475	12/04		Derby	Leeds	10/26
2584	3/00	476	2/05		Derby	Leeds	8/27
2585	3/00	477	12/04	7/22	Derby	Kentish Town	2/51
2586	4/00	478	4/05	12/22	Derby	Nottingham	10/50
2587	4/00	479	3/05	3/17	Derby	Western Div	12/49
2588	4/00	480	7/05	6/22	Derby	Normanton	2/54
2589	5/00	481	7/05		Derby	Leeds	4/26
2590	5/00	482	8/05	9/14	Derby	Millhouses	7/57

The 1667 replacements

Dimensions

Cylinders (2 inside)	19in x 26in
Coupled wheel diameter	7ft 0in
Bogie wheel diameter	3ft 6in
Stephenson valve gear	
Boiler pressure	'D' type, 160lb psi
Heating surface	1,205sq ft
Grate area	19.6sq ft
Axleweight	16 tons 19 cwt
Weight: Engine	47 tons 4 cwt
Tender	37 tons
Total	84 tons 4 cwt
Water capacity	3,250 gallons
Coal capacity	4 tons
Tractive effort	14,440lbs

Rebuilt as superheated 483 class

Cylinder diameter	20½in x 26in
Piston valves	8in diameter
Coupled wheel diameter	7ft 0½in
Bogie wheel diameter	3ft 6½in
Boiler pressure	160lbs psi
Heating surface	1,483sq ft (of which superheater was 313sq ft) (Later 1,410sq ft with superheater 253sq ft)
Grate area	21.1sq ft
Axleload	17½ tons
Weight Engine	53 tons 7 cwt
Tender	39 tons 16 cwt
Total	93 tons 3 cwt
Water capacity	3,250 gallons
Coal capacity	4 tons
Tractive effort	16,551lbs

Weight diagram

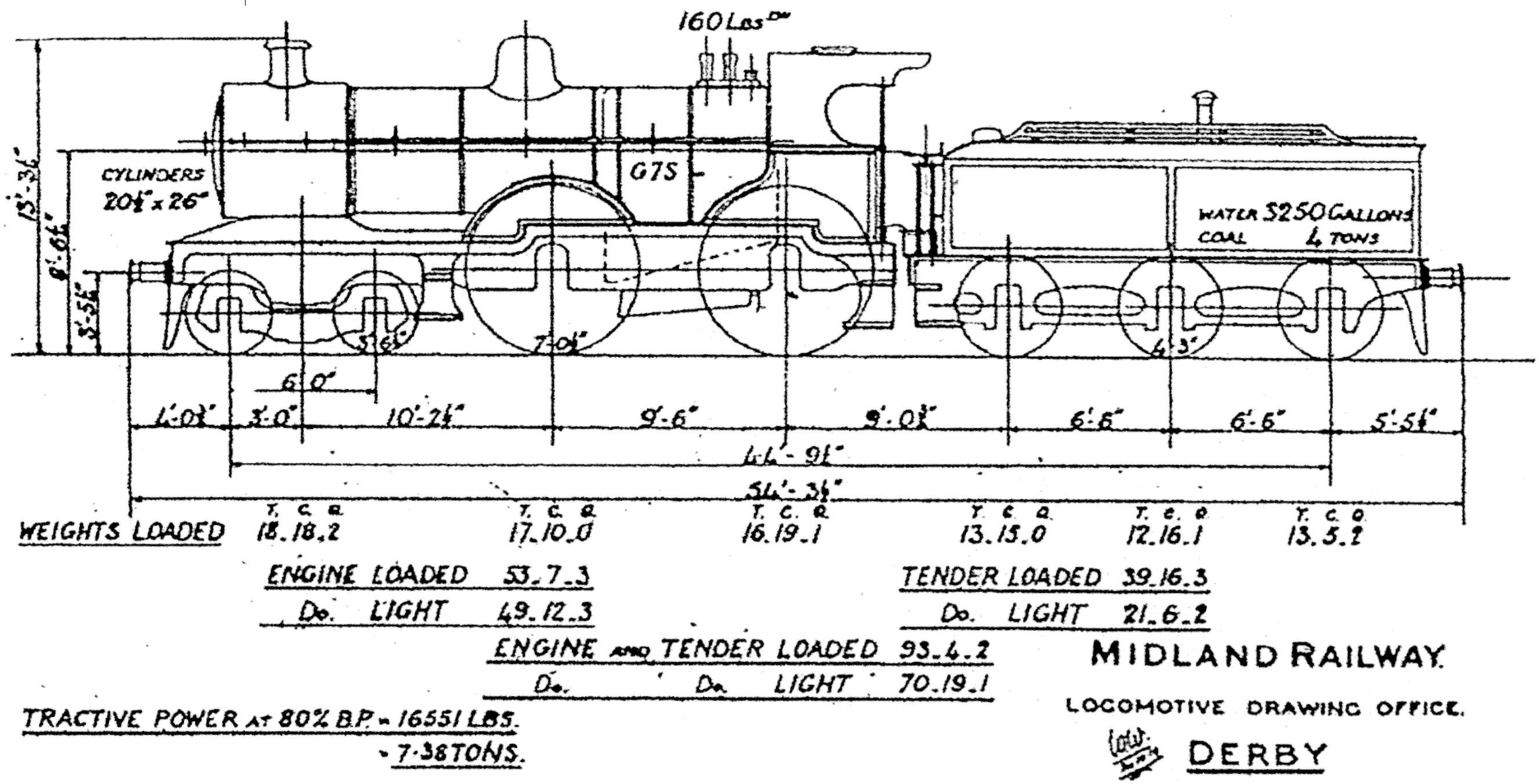

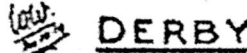

Statistics

No.	Built	1907	H boiler	483 class	First depot	Last depot	Withdrawn
1667	3/97	483	7/07	11/12	Belle Vue		9/49
1668	12/96	484	7/07	5/13	Belle Vue	Skipton	8/53
1669	9/98	485	6/07	5/13	Belle Vue	Leicester	7/57
1670	4/01	486	1/08	10/12	Nottingham	Bristol	2/57
1671	10/98	487	2/07	9/12	Nottingham	Nottingham	1/61
1672	10/96	488	2/08	3/12	Nottingham	Lancaster	12/50
1673	5/01	489	3/08	4/13	Nottingham	Gloucester	9/60
1674	6/01	490	9/06	6/12	Nottingham		1/50
1675	10/96	491	10/06	5/13	Kentish Town	Leeds	9/60
1676	3/97	492	3/08	9/12	Kentish Town		12/48

The 150 series

Dimensions

Cylinders (2 inside)	19in x 26in
Coupled wheel diameter	7ft 0in
Bogie wheel diameter	3ft 6in
Stephenson valve gear	
Boiler pressure	'D' type, 160lb psi
Heating surface	1,205sq ft
Grate area	19.6sq ft
Axleweight	16 tons 19 cwt
Weight: Engine	47 tons 4 cwt
Tender	37 tons
Total	84 tons 4 cwt
Water capacity	3,250 gallons
Coal capacity	4 tons

Statistics

No.	Built	1907	H boiler	483 boiler	First depot	Last depot	Withdrawn
150	9/97	493	9/06	8/12	Leicester	Nottingham	6/59
153	9/97	494	10/07	2/12	Leicester		12/48
154	10/97	495	6/07	3/12	Leicester	Rhyl	7/57
155	10/97	496	6/07	12/12	Leicester		2/49
204	10/97	497	6/07	11/12	Saltley	Spital Bridge	10/51
205	11/97	498	8/06	3/13	Trafford Park		9/50
206	11/97	499	9/07	2/13	Trafford Park	Rowsley	9/52
207	12/97	500	10/06	10/12	Trafford Park		7/49
208	12/97	501	8/07	9/12	Trafford Park	Bristol	7/60
209	12/97	502	8/06	5/12	Trafford Park	Nottingham	2/61

The 2421 series

Dimensions

Cylinders (2 inside)	18½in x 26in
Coupled wheel diameter	7ft 0in
Bogie wheel diameter	3ft 6in
Stephenson valve gear	
Boiler pressure	'D' type, 170lb psi
Heating surface	1,205sq ft
Grate area	19.6sq ft
Axleweight	16 tons 19 cwt
Weight: Engine	47 tons 4 cwt
Tender	37 tons
Total	84 tons 4 cwt
Water capacity	3,250 gallons
Coal capacity	4 tons

Statistics

No.	Built	1907	H boiler	483 boiler	First depot	Last depot	Withdrawn
2421	10/99	503	12/06	6/12	Nottingham	Hasland	11/52
2422	10/99	504	2/08	9/12	Nottingham	Nottingham	1/61
2423	10/99	505	4/08	5/12	Nottingham	Templecombe	10/53
2424	10/99	506	12/06	10/12	Nottingham	Toton	11/49
2425	10/99	507	12/06	10/12	Nottingham	Stafford	6/52
2426	11/99	508	6/07	3/12	Nottingham	Nuneaton	7/51
2427	11/99	509	12/06	5/12	Nottingham	Bath Green Pk	6/57
2428	11/99	510	4/08	11/12	Nottingham	Normanton	5/49
2429	11/99	511	9/07	5/13	Nottingham	Toton	1/61
2430	11/99	512	2/08	8/12	Nottingham	Saltley	1/50
2431	12/99	513	5/08	11/12	Nottingham	Nottingham	10/59
2432	12/99	514	3/08	12/12	Nottingham	Leeds Holbeck	1/52
2433	12/99	515	11/07	6/12	Nottingham	Rhyl	8/50
2434	12/99	516	5/08	10/12	Nottingham	Derby	3/50
2435	12/99	517	2/08	2/13	Nottingham	Bourneville	6/49
2436	12/99	518	11/07	11/12	Leeds	Leeds Holbeck	9/56
2437	12/99	519	7/07	5/13	Leeds	Burton	12/57
2438	12/99	520	9/06	5/12	Leeds	Royston	3/56
2439	12/99	521	2/07	2/13	Leeds	Royston	10/57
2440	12/99	522	8/07	5/12	Leeds	Crewe North	10/55

The 60 series

Dimensions

Cylinders	19½in x 26in (60-66, 93, 138, 139)
	19in x 26in remainder (most were later lined up to 18 ½ in)
Coupled wheel diameter	7ft 0½in
Bogie wheel diameter	3ft 6½in
Piston valves	8in diameter
Boiler pressure	E boiler, 170lbs psi
Heating surface	1,205sq ft
Grate area	19.6sq ft
Axleload	16 tons 19 cwt
Weight – Engine	47 tons 4 cwt
– Tender	38 tons
– Total	85 tons 4 cwt
Water capacity	3,500 gallons
Coal capacity	4 tons
Tractive effort	14,440lbs

Weight diagram

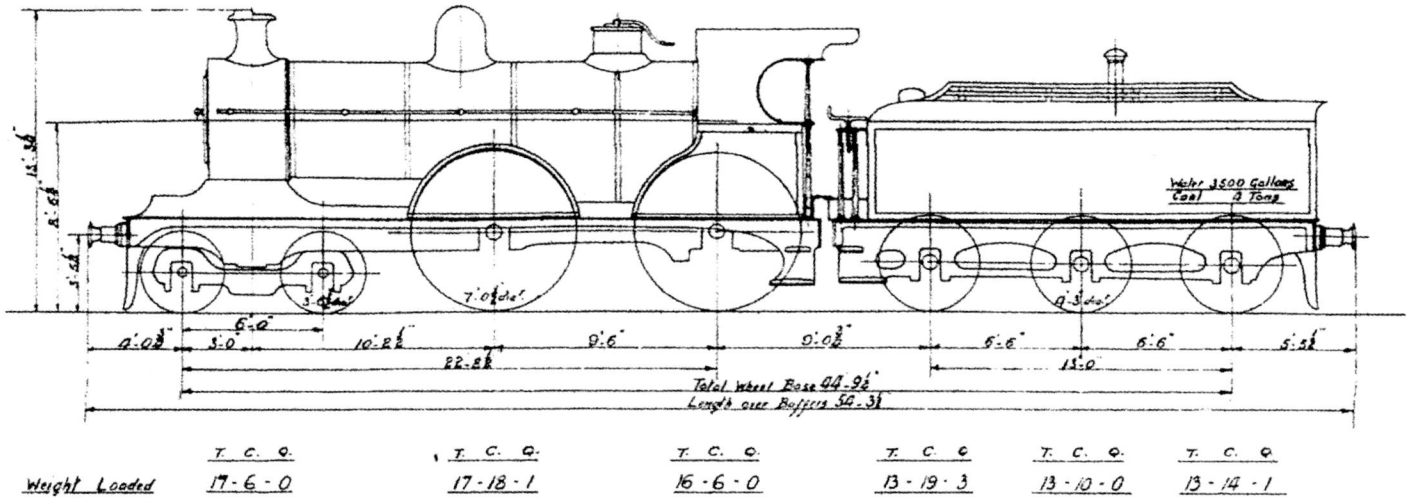

Statistics

No.	Built	1907	H boiler	483 boiler	First depot	Last depot	Withdrawn
60	6/98	523	8/06	3/15	Leicester	Gloucester	10/52
61	7/98	524	6/06	4/14	Leicester	Preston	7/54
62	8/98	525	5/07	8/13	Leicester	Burton	5/57
63	8/98	526	5/06	11/13	Leicester	Burton	7/56
64	9/98	527	5/06	11/13	Leicester	Bath	3/56

No.	Built	1907	H boiler	483 boiler	First depot	Last depot	Withdrawn
65	10/98	528	10/06	5/13	Leicester	Nuneaton	12/52
66	10/98	529	8/06	9/13	Leicester	Crewe	5/54
93	11/98	530	3/07	5/13	Bristol	Gloucester	1/51
138	11/98	531	3/07	5/14	Bristol	Walsall	9/56
139	11/98	532	10/06	8/13	Bristol	Spital Bridge	1/52
67	6/99	533	3/07	6/13	Bristol		1/50
68	7/99	534	11/06	2/13	Bristol	Nottingham	6/59
69	7/99	535	11/06	6/13	Bristol	Nottingham	9/55
151	8/99	536	11/07	6/13	Derby	Derby	5/59
152	8/99	537	10/06	11/13	Derby	Templecombe	9/62
165	9/99	538	4/07	1/14	Derby	Derby	5/59
166	10/99	539	1/07	12/13	Derby	Longsight	8/54
167	10/99	540	4/07	4/14	Derby	Gloucester	2/62
168	10/99	541	7/07	10/13	Derby	Bourneville	3/58
169	10/99	542	5/08	10/13	Derby	Nottingham	9/59
805	3/01	543	7/06	9/13	Carlisle	Leicester	1/61
806	4/01	544	8/06	6/14	Carlisle		11/49
807	4/01	545	6/06	1/14	Carlisle		12/48
808	5/01	546	6/06	10/13	Carlisle	Nottingham	6/51
809	5/01	547	6/06	5/14	Carlisle	Kentish Town	2/53
2636	6/01	548	6/07	7/13	Leeds	Kentish Town	1/61
2637	6/01	549	4/07	6/13	Leeds	Millhouses	4/51
2638	6/01	550	5/07	1/15	Leeds	Nottingham	6/59
2639	7/01	551	6/07	10/13	Leeds	Bedford	10/53
2640	7/01	552	9/06	10/13	Leeds	Leeds	6/60
2591	5/01	553	5/07	3/14	Leicester	Nottingham	10/58
2592	5/01	554	2/08	12/13	Kentish Town		9/49
2593	5/01	555	1/07	6/13	Kentish Town		4/49
2594	5/01	556	5/06	6/13	Leicester	Hasland	4/56
2595	5/01	557	3/07	3/14	Leicester	Nottingham	3/61
2596	5/01	558	2/07	9/13	Kentish Town	Spital Bridge	3/52
2597	6/01	559	6/06	12/14	Kentish Town	Chester	12/57
2598	6/01	560	11/07	8/13	Kentish Town	Nottingham	12/52
2599	6/01	561	7/07	7/13	Kentish Town		3/50
2600	6/01	562	2/07	11/13	Leicester	Manningham	11/55

The Somerset and Dorset 4-4-0s
Dimensions

The small 4-4-0s of 1891-6
Cylinders (2 inside)	18in x 24in
Coupled wheel diameter	5ft 9in
Bogie wheel diameter	3ft 0in
Boiler pressure	150lbs psi
Heating surface	1,202sq ft
Grate area	16sq ft
Axleload	15 tons 2 cwt
Weight – Engine	39 tons
– Tender	29 tons 18 cwt
– Total	68 tons 18 cwt
Water capacity	2,200 gallons
Coal capacity	3 tons

The 4-4-0s of 1903
Cylinders (2 inside)	18in x 26in
Coupled wheel diameter	6ft 0in
Bogie wheel diameter	3ft 1in
Boiler pressure	175lbs psi
Heating area	1,420sq ft
Grate area	21.1sq ft
Axleload	16 tons 9 cwt
Weight – Engine	46 tons 4 cwt
– Tender	35 tons 2 cwt
– Total	81 tons 6 cwt
Water capacity	2,950 gallons
Coal capacity	3 tons

The 4-4-0s of 1908
As above apart from:
Heating surface	1,347sq ft
Axleload	16 tons 12 cwt
Weight – Engine	47 tons 8 cwt
– Tender	36 tons 13 cwt

The superheated 4-4-0s of 1914 and 1921
Cylinders (2 inside)	20½in x 26in
Coupled wheel diameter	7ft 0½in
Bogie wheel diameter	3ft 6½in
Boiler pressure	160lbs psi
Heating surface	1,170sq ft
Superheating surface	313sq ft
Grate area	21.1sq ft
Axleload	17½ tons
Weight – Engine	53 tons 7 cwt
– Tender	37 tons
– Total	90 tons 7 cwt
Water capacity	3,250 gallons
Coal capacity	4 tons

Statistics

No.	Built	Renumbered	Reboilered	Last depot	Withdrawn	
15	5/91		4/05	*	8/28	
16	5/91		4/06	*	8/28	
17	5/91		8/04**	*	6/31	LMS 302
18	5/91	15 (8/28)	12/04**	*	9/31	LMS 301
67	1/96		12/07	*	8/20	Replaced
68	1/96		5/08	*	11/21	Replaced
14	2/97		12/10	*	1/30	
45	2/97	18 (8/28)	8/09**	*	2/32	LMS 303
69	11/03			*	4/21	Replaced
70	11/03			*	4/14	Replaced
71	11/03			*	5/14	Replaced
77	3/08	320 (1930)	5/26	*	9/31	
78	3/08	321 (1930)	11/21	Burton	3/38	
70	5/14	39 (8/28)		Stafford	2/53	322 (1930), 40322 (1948)
71	4/14	40 (8/28)		Leeds	9/56	323 (1930), 40323 (1948)
67	4/21	41 (8/28)		Rhyl	1/53	324 (1930), 40324 (1948)
68	4/21	42 (8/28)		Burton	10/51	325 (1930), 40325 (1948)
69	4/21	43 (8/28)		Derby	5/56	326 (1930), 40326 (1948)
44	6/28	633 (1930)		Burton	10/59	40633 (1948)
45	6/28	634 (1930)		Templecombe	1962	40634 (1948)
46	7/28	635 (1930)		Llandudno Jn		40635 (1948)

* All remained at S&D depots throughout their lives
** Belpaire boiler in 1925.

The Midland & Great Northern 4-4-0s
Dimensions
D52

Cylinders (2 inside)	18½in x 26in
Coupled wheel diameter	6ft 6½in
Bogie wheel diameter	3ft 3½in
Boiler pressure	160lbs psi
Heating surface	1,078sq ft
Grate area	17.5sq ft
Axleload	16 tons
Weight – Engine	42 ton 18 cwt
– Tender	33 tons 11 cwt
– Total	76 tons 9 cwt

Water capacity　　　　　2,950 tons
Coal capacity　　　　　　3 tons
Tractive effort　　　　　15,416lbs

D53
As above apart from:
Weight – Engine　　　　44 tons 7 cwt
Axleload　　　　　　　　16 tons 10 cwt

D54
As D52 above apart from:
Boiler pressure　　　　　175lbs psi
Heating surface　　　　　1,384sq ft
Grate area　　　　　　　21sq ft
Axleload　　　　　　　　17 tons
Weight – Engine　　　　49 tons 18 cwt
Tractive effort　　　　　16,862lbs

Statistics

No.	Built	Renumbered	Rebuilt	Last depot	Withdrawn	
36	5/94		D53 5/29	Peterborough East	1/37	
37	5/94			Melton Constable	2/37	
38	5/94	038 (12/37)		Melton Constable	9/43	
39	5/94		D54 1/24	Melton Constable	2/37	H boiler 1908
42	5/94	042 (11/36)		New England	6/40	
43	5/94	043 (9/37)		Melton Constable	6/43	
44	6/94	044 (7/37)	D53 5/30	Peterborough East	8/41	
45	6/94		D54 1909	South Lynn	11/36	
46	6/94	046 (3/37)	D54 1915	Yarmouth Beach	3/43	
47	6/94	047 (7/37)		Melton Constable	6/42	
48	7/94			Melton Constable	11/37	
49	7/94	049 (10/37)	D53 2/31	New England	9/41	
50	7/94	050 (9/37)	D53 11/29	Yarmouth Beach	1/45	Allocated 2052 (1946)
1	8/94	01 (10/36)		Yarmouth Beach	11/37	
2	8/94	02 (1/37)	D53 4/31	Melton Constable	5/43	
3	8/94			Yarmouth Beach	6/37	
4	8/94			Melton Constable	2/38	
5	8/94	05 (11/36)		Melton Constable	7/37	
6	8/94	06 (10/37)	D53 8/30	Yarmouth Beach	3/44	Allocated 2053 (1946)
7	8/94	07 (10/36)		South Lynn	6/37	

No.	Built	Renumbered	Rebuilt	Last depot	Withdrawn	
11	9/94	011 (9/37)		Peterborough East	8/42	
12	11/94	012 (11/37)		Yarmouth Beach	8/42	
13	11/94	013 (5/37)		New England	9/41	
14	11/94			Melton Constable	2/37	
17	11/94			Melton Constable	10/37	
18	11/94			South Lynn	2/37	
51	8/96	051 (9/37)	D54 1915	South Lynn	5/43	
52	8/96	052 (10/36)	D54 1913	South Lynn	2/43	
53	8/96	053 (7/37)	D54 1910	South Lynn	1/40	
54	9/96	054 (1/37)	D54 1914	Melton Constable	10/39	
55	9/96	055 (4/37)	D54 1/25	Melton Constable	11/43	H boiler 1908
56	9/96	056 (2/37)	D54 1912	Melton Constable	11/43	
57	9/96		D54 1912	South Lynn	2/37	
74	10/99			South Lynn	5/37	
75	10/99			South Lynn	2/37	
76	10/99	076 (11/37)		New England	7/43	
77	11/99	077 (10/37)	D53 12/30	Yarmouth Beach	1/45	Allocated 2054 (1946)
78	11/99	078 (11/36)		South Lynn	2/38	
79	11/99	079 (11/36)		Peterborough East	2/37	
80	11/99			Melton Constable	2/37	

* All remained on the former M&GN area at Lynn or Melton Constable

Midland Class 3 'Belpaires'
Dimensions
2606 series

Cylinders	19½in x 26in
Coupled wheel diameter	6ft 9in
Bogie wheel diameter	3ft 6½in
Boiler pressure	175lbs psi (180 later)
Heating surface	1,519sq ft
Grate area	25sq ft
Axleload	18 tons 5 cwt
Weight – Engine	51 tons 12 cwt
– Tender	53 tons 11 cwt (later 39 tons 3 cwt)
– Total	105 tons 3 cwt
Water capacity	4,000 gallons (later 3,500 gallons)
Coal capacity	5 tons (later 4 tons)
Tractive effort	18,675lbs

2781 series

As above, except:
Boiler pressure	180lbs psi
Heating surface	1,528sq ft
Weight – Engine	53½ tons
– Tender	38¾ tons
– Total	92¼ tons

860 series

As above, except
Boiler pressure	200lbs psi
Heating surface	1,455sq ft
Tractive effort	20,750lbs

Superheated

As above, except
Cylinders	20½in x 26in
Boiler pressure	175lbs psi
Heating surface	1,496sq ft
Weight – Engine	51 tons 8 cwt
– Tender	40 tons 15 cw
– Total	92 tons 3 cwt
Tractive effort	20,065lbs

Weight diagram

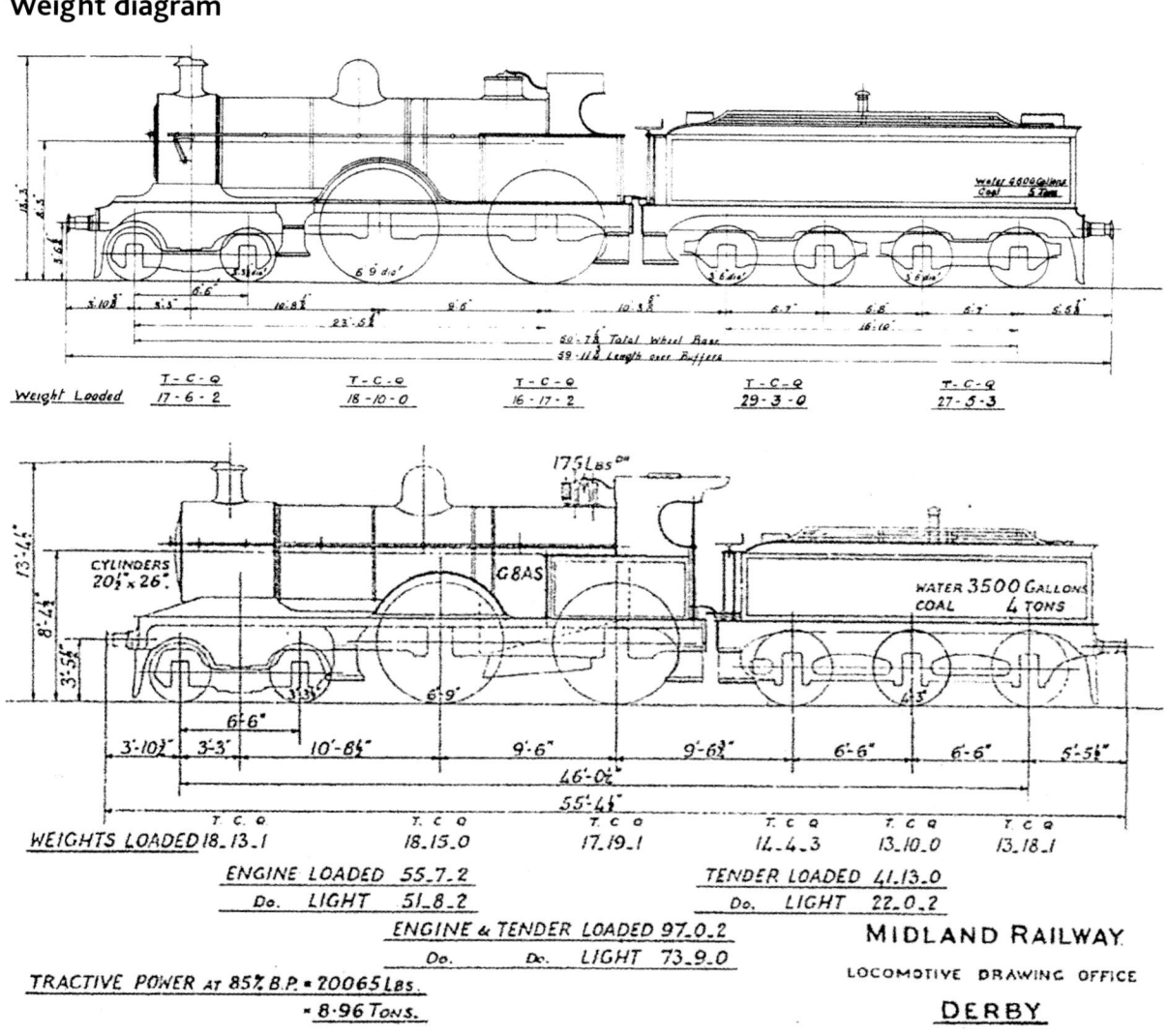

Statistics

No.	Built	1907	Reboilered	Superheated	First depot	Last depot	Withdrawn
2606	9/00	700		9/13	Leeds	Leicester	11/27
2607	10/00	701		12/13	Leeds	Leicester	8/31
2608	10/00	702		6/14	Leeds	Leicester	11/31
2609	10/00	703		10/13	Leeds	Leicester	5/28
2610	11/00	704		5/14	Leeds	Leicester	8/27
800	1/01	705		2/14	Kentish Town	Manchester	8/27
801	1/01	706		2/21	Kentish Town	Manchester	5/32
802	2/01	707	7/10 (GXA)	5/25	Carlisle	Leicester	12/47
803	3/01	708		11/14	Carlisle	Manchester	2/36
804	6/01	709		4/14	Carlisle	Kentish Town	7/28
2781	1/02	710	8/17 (G8A)	6/20	Leeds	Leicester	4/41
2782	1/02	711	9/11 (G8AX)	3/22	Leeds	Rowsley	3/49
2783	1/02	712		11/23	Leeds	Saltley	12/38
2784	1/02	713		10/19	Manchester	Saltley	11/35
2785	2/02	714	11/13 (G8A)	12/22	Manchester	Saltley	12/28
2786	2/02	715		6/21	Manchester	Gloucester	1/48
2787	4/02	716		12/23	Kentish Town	Sheffield	12/47
2788	4/02	717		6/19	Kentish Town	Gloucester	6/35
2789	5/02	718		5/20	Kentish Town	Bristol	4/36
2790	6/02	719	1/14 (G8A)	2/21	Kentish Town	Nottingham	10/47
810	10/02	720	2/20 (G8A)	3/25	Carlisle	Bedford	7/49
811	10/02	721	9/10 (G8AX) 5/18 (G8)	4/21	Carlisle	Peterborough	4/45
812	11/02	722		2/17	Leeds	Nottingham	6/39
813	11/02	723		6/16	Leeds	Nottingham	6/39
814	11/02	724		6/17	Leeds	Leeds	4/36
815	12/02	725		12/17	Leeds	Bedford	10/47
816	12/02	726		11/17	Leeds	Canklow	9/52 (40726)
817	12/02	727		12/21	Leeds	Canklow	6/50
818	12/02	728		5/22	Leeds	Sheffield	7/52 (40728)
819	12/02	729		7/16	Leeds	Sheffield	6/51
820	4/03	730		6/21	Leeds	Sheffield	7/36
821	5/03	731		6/22	Leeds	Sheffield	12/48
822	5/03	732		11/17	Leeds	Leeds	5/29
823	6/03	733		9/21	Manchester	Nottingham	10/36

Dimensions, Weight Diagrams & Statistics • 325

No.	Built	1907	Reboilered	Superheated	First depot	Last depot	Withdrawn
824	6/03	734	12/18 (G8A)	9/23	Manchester	Lancaster	10/49
825	6/03	735		2/20	Manchester	Lancaster	2/49
826	6/03	736		6/21	Manchester	Lancaster	2/48
827	7/03	737			Manchester	Leeds	9/25
828	7/03	738		9/22	Manchester	Rowsley	12/47
829	9/03	739		5/20	Manchester	Nottingham	12/49
830	12/03	740	6/15 (G8A)	6/22	Manchester	Rowsley	11/49 (40740)
831	12/03	741		6/20	Manchester	Bristol	10/51 (40741)
832	12/03	742	7/23 (G8A)		Manchester	Leeds	3/25
833	12/03	743	12/14 (G8A)	5/21	Manchester	Bedford	7/52 (40743)
834	12/03	744		2/20	Manchester	Manchester	8/36
835	1/04	745		5/21	Derby	Derby	3/50 (40745)
836	2/04	746		1/20	Derby	Manchester	12/35
837	2/04	747	1/22 (G8A)	9/24	Derby	Leeds	6/51 (40747)
838	2/04	748		8/22	Derby	Derby	7/48
839	2/04	749	2/13 (G8A)		Derby	Manchester	5/25
840	6/04	750		11/23	Manchester	Bedford	6/39
841	7/04	751	2/14 (G8A)		Manchester	Gloucester	12/26
842	8/04	752		12/22	Manchester	Gloucester	4/29
843	9/04	753	9/14 (G8A)	5/24	Manchester	Bedford	6/35
844	9/04	754		12/17	Manchester	Bedford	2/30
845	10/04	755	6/14 (G8A)	9/22	Manchester	Bedford	7/46
846	10/04	756	4/11 (G8AX) 1/21 (G8)	2/23	Manchester	Peterborough	5/49
847	11/04	757		3/21	Manchester	Nottingham	5/48
848	11/04	758	10/16 (G8A)	4/25	Manchester	Leeds	3/51 (40758)
849	1/05	759	6/21 (G8A)	5/23	Manchester	Derby	2/46
850	2/05	760		9/21	Derby	Bedford	4/46
851	2/05	761		4/20	Derby	Derby	4/29
852	3/05	762		3/23	Derby	Bedford	2/51
853	3/05	763		11/23	Derby	Bedford	12/47
854	3/05	764		9/23	Derby	Bedford	12/34
855	3/05	765		2/19	Derby	Sheffield	10/47
856	3/05	766		1/17	Derby	Derby	5/28
857	3/05	767		9/21	Derby	Peterborough	5/47
858	4/05	768		1/19	Derby	Bedford	6/36

No.	Built	1907	Reboilered	Superheated	First depot	Last depot	Withdrawn
859	4/05	769		11/16	Derby	Derby	10/36
860	7/05	770		11/20	Manchester	Derby	4/36
861	7/05	771		7/23	Manchester	Derby	7/35
862	7/05	772			Manchester	Bedford	6/26
863	7/05	773		11/19	Manchester	Rowsley	4/40
864	7/05	774		3/23	Manchester	Nottingham	4/40
865	8/05	775		5/23	Manchester	Nottingham	1/47
866	8/05	776		7/20	Manchester	Manchester	6/36
867	8/05	777		6/24	Manchester	Manchester	6/39
868	8/05	778			Manchester	Manchester	5/25
869	9/05	779			Manchester	Manchester	10/25

Midland Class 4 Compound

Dimensions

Cylinder diameter – high pressure	19in x 26in
– low pressure	21in x 26in
Stephenson's Link Motion	
Coupled wheel diameter	7ft 0in
Bogie wheel diameter	3ft 6½in
Boiler pressure	195lbs psi
Heating surface	1,598sq ft
Grate area	26sq ft
Axleload	19½ tons
Weight – Engine	59 tons 10 cwt
– Tender	52 tons 12 cwt
– Total	112 tons 2 cwt
Water capacity	4,500 gallons
Coal capacity	5 tons
Tractive effort	19,110lbs

Altered dimensions after superheating

Boiler pressure	200lbs psi
Grate area	28.4sq ft
Heating surface	1,681sq ft (incl superheater 360sq ft)
Axleload	19¾ tons
Weight – Engine	61 tons 14 cwt
– Tender	45 tons 18 cwt
– Total	107 tons 12 cwt
Water capacity	3,500 gallons
Coal capacity	7 tons
Tractive effort	21,840lbs

Dimensions, Weight Diagrams & Statistics

Weight diagrams

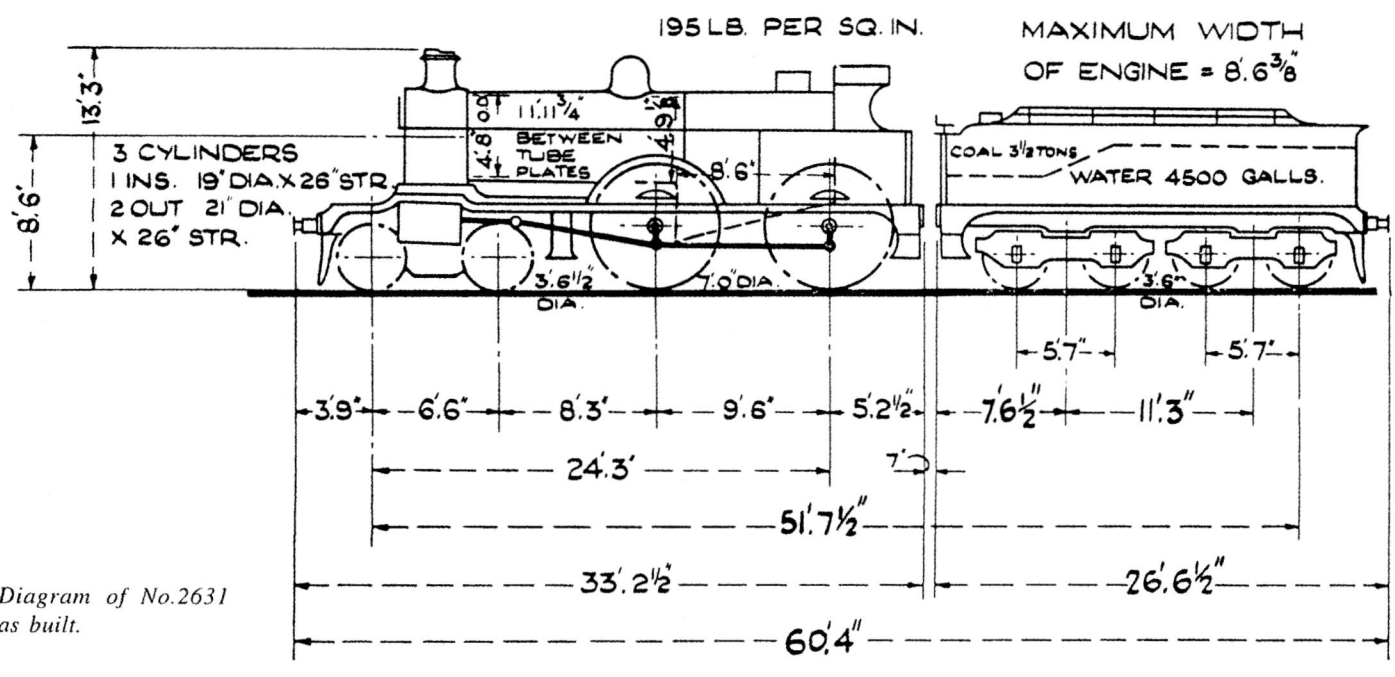

Diagram of No.2631 as built.

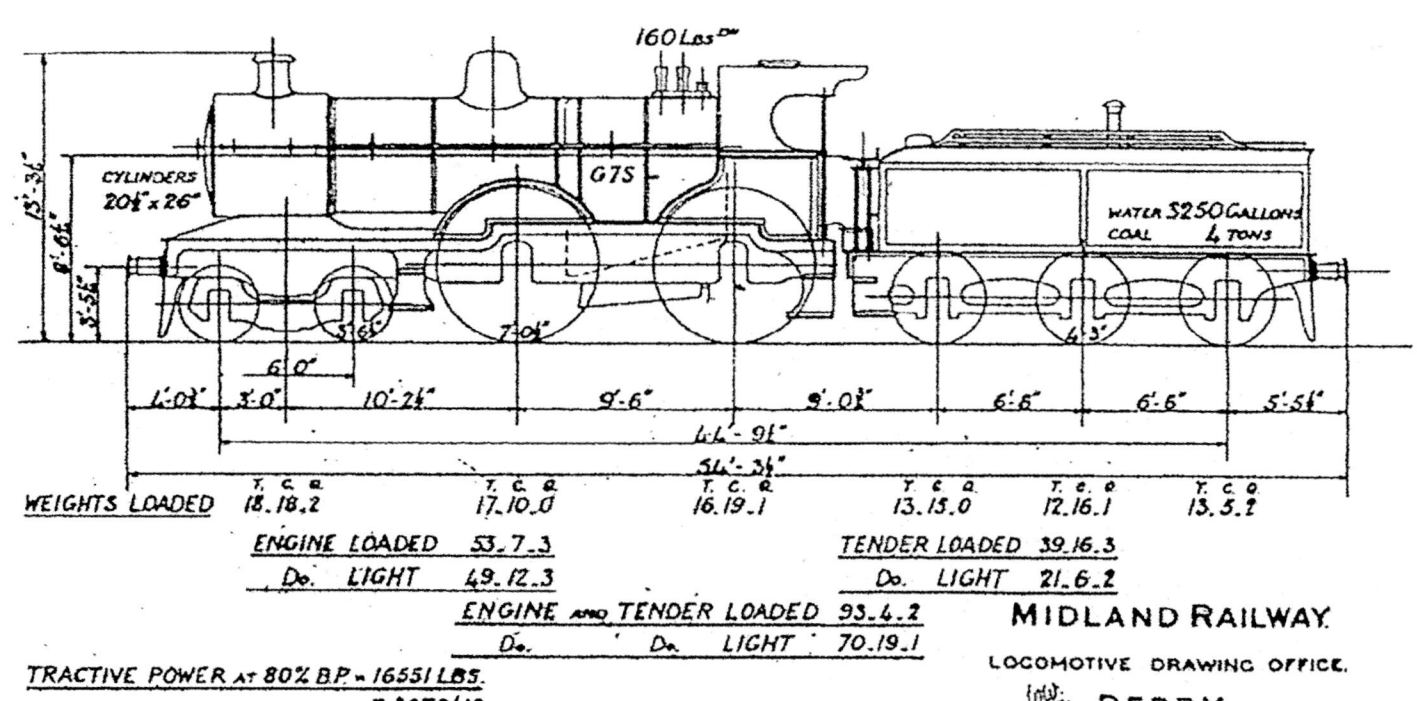

Statistics

No.	Built	1907	BR	Superheated	First depot	Last depot	Withdrawn	
2631	1/02	1000	41000	11/14	Leeds	Derby	10/51	Preserved
2632	1/02	1001	41001	1/15	Carlisle	Gloucester	11/51	
2633	7/03	1002		1/19	Nottingham	Gloucester	6/48	
2634	9/03	1003	41003	5/15	Kentish Town	Derby	4/51	
2635	11/03	1004	41004	11/14	Kentish Town	Manningham	2/52	
1000	10/05	1005	41005	3/23	Kentish Town	Lancaster	7/51	
1001	11/05	1006	41006	8/22	Kentish Town	Leicester	5/51	
1002	11/05	1007	41007	6/22	Kentish Town	Nottingham	5/52	
1003	11/05	1008		7/19	Kentish Town	Leicester	7/49	
1004	12/05	1009	41009	11/22	Kentish Town	Bedford	12/51	
1005	12/05	1010		1/27	Kentish Town	Leeds	7/49	
1006	12/05	1011	41011	12/22	Kentish Town	Leicester	2/51	
1007	12/05	1012	41012	7/21	Kentish Town	Bristol	1/51	
1008	12/05	1013		3/25	Kentish Town	Leeds	6/49	
1009	12/05	1014	41014	5/19	Kentish Town	Derby	5/52	
1010	3/06	1015	41015	5/21	Kentish Town	Nottingham	12/51	
1011	3/06	1016	41016	5/22	Nottingham	Millhouses	11/51	
1012	3/06	1017	41017	1/26	Nottingham	Leeds	8/50	
1013	4/06	1018		12/26	Nottingham	Leeds	7/48	
1014	4/06	1019	41019	4/24	Nottingham	Nottingham	12/51	
1015	4/06	1020	41020	2/22	Nottingham	Kentish Town	5/51	
1016	4/06	1021	41021	9/27	Nottingham	Nottingham	10/52	
1017	5/06	1022	41022	1/28	Nottingham	Lancaster	4/50	
1018	5/06	1023	41023	12/22	Nottingham	Derby	8/51	
1019	5/06	1024		3/22	Nottingham	Nottingham	10/48	
1020	5/06	1025	41025	10/24	Nottingham	Gloucester	1/53	
1021	6/06	1026		10/27	Nottingham	Manchester	9/48	
1022	6/06	1027		8/26	Leeds	Leeds	10/48	
1023	6/06	1028	41028	11/23	Leeds	Bristol	10/52	
1024	7/06	1029		5/24	Leeds	Bristol	6/48	
1025	9/06	1030	41030	4/25	Leeds	Bristol	9/51	
1026	9/06	1031		11/22	Leeds	Bedford	12/49	
1027	10/06	1032	41032	1/26	Leeds	Nottingham	3/52	
1028	10/06	1033		10/23	Leeds	Bristol	9/48	
1029	12/06	1034	41034	1/23	Leeds	Bedford	5/50	
1035	11/08		41035	6/22	Leeds	Bourneville	5/52	
1036	11/08			11/23	Leeds	Carnforth	10/48	

No.	Built	1907	BR	Superheated	First depot	Last depot	Withdrawn
1037	12/08		41037	2/21	Manchester	Millhouses	3/51
1038	12/08		41038	11/22	Manchester	Bedford	8/52
1039	12/08		41039	9/20	Manchester	Gloucester	6/50
1040	1/09		41040	7/13	Manchester	Leeds	5/52
1041	1/09		41041	5/23	Manchester	Leicester	12/51
1042	2/09			6/20	Kentish Town	Leicester	6/49
1043	3/09		41043	12/25	Kentish Town	Derby	12/51
1044	3/09		41044	12/26	Kentish Town	Bedford	10/52

Midland Class 4 990 class

Dimensions

Cylinder diameter	19in x 26in
Walschaerts valve gear (Deeley 'scissors' version)	
Coupled wheel diameter	6ft 6½in
Bogie wheel diameter	3ft 3½in
Boiler pressure	200lbs psi
Grate area	28.4sq ft
Heating surface	1,557.4sq ft
Axleload	19¾ tons
Weight Engine	58 tons 10 cwt
Tender	45 tons 18 cwt
Total	104 tons 8 cwt
Water capacity	3,500 gallons
Coal capacity	7 tons
Tractive effort	23,662lbs

Weight diagram

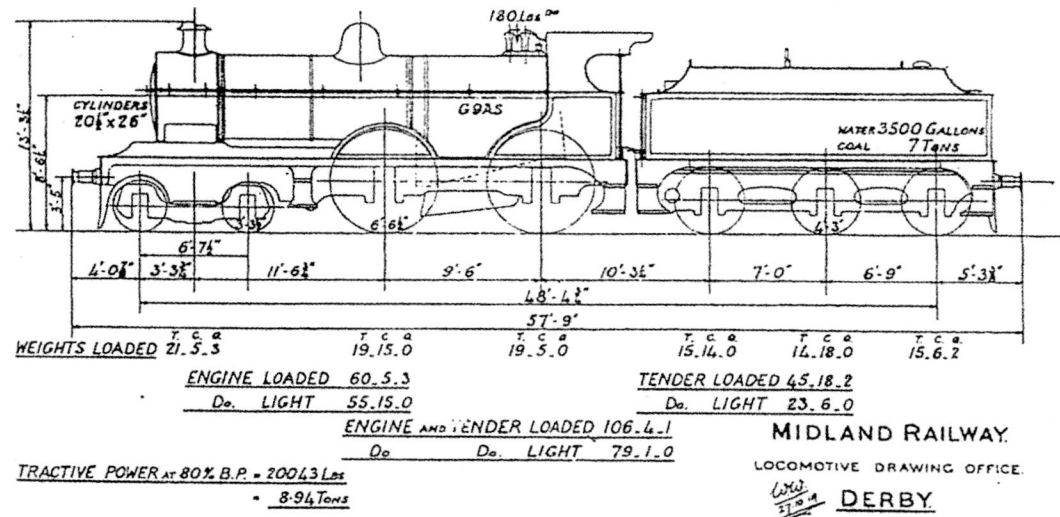

Statistics

No.	Built	1926	Superheated	First depot	Last depot	Withdrawn
990	4/09		3/13	Kentish Town	Carlisle	3/25
991	5/09	801	1/13	Kentish Town	Carlisle	12/27
992	6/09		8/12	Kentish Town	Carlisle	3/28
993	6/09	803	7/12	Manchester	Carlisle	1/28
994	6/09		9/13	Manchester	Carlisle	7/26
995	7/09	805	3/12	Manchester	Carlisle	1/28
996	9/09	806	1/14	Leeds	Carlisle	10/28
997	9/09	807	7/13	Leeds	Carlisle	12/27
998	10/09	808	5/10	Leeds	Carlisle	12/28
999	3/07	809	5/11	Derby	Carlisle	12/28

LMS 4P Compound

Dimensions

Cylinder diameter – high pressure	19¾ in x 26in (adjusted to 19in)
– low pressure	21¾ in x 26in (adjusted to 21in)
Stephenson's Link Motion	
Coupled wheel diameter	6ft 9in
Bogie wheel diameter	3ft 6½in
Boiler pressure	200lbs psi
Heating surface	1,607.7sq ft (of which superheater 290.7sq ft)
Grate area	28.4sq ft
Axleload	19¾ tons
Weight – Engine	61 tons 14 cwt
– Tender	42 tons 14 cwt
– Total	104 tons 8 cwt
Water capacity	3,500 gallons
Coal capacity	5½ tons
Tractive effort	22,649lbs

Weight diagram

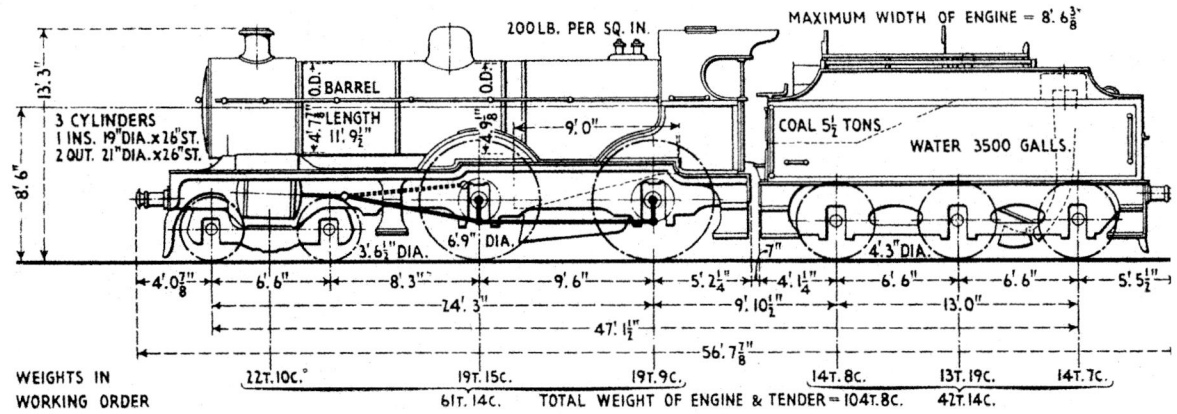

Statistics

No.	Built	BR	First depot	1933 Allocation	Last depot	Withdrawn
1045	2/24	41045	Midland Div	Derby	Lancaster	6/57
1046	2/24	41046	Midland Div	Derby	Bourneville	1/53
1047	2/24	41047	Midland Div	Derby	Gloucester	2/54
1048	3/24	41048	Midland Div	Derby	Wellingborough	11/57
1049	3/24	41049	Midland Div	Derby	Derby	3/59
1050	3/24	41050	Midland Div	Derby	Derby	6/56
1051	3/24	41051	Camden	Kentish Town	Kentish Town	11/54
1052	4/24	41052	Camden	Kentish Town	Trafford Park	4/53
1053	4/24	41053	Camden	Kentish Town	Leicester	6/56
1054	4/24	41054	Camden	Kentish Town	Bedford	10/54
1055	4/24	41055	Camden	Kentish Town	Llandudno Jcn	3/53
1056	5/24	41056	Camden	Millhouses	Leeds	11/53
1057	5/24	41057	Midland Div	Derby	Manningham	5/53
1058	5/24	41058	Midland Div	Derby	Millhouses	2/54
1059	6/24	41059	Midland Div	Derby	Bedford	12/55
1060	6/24	41060	Midland Div	Derby	Lancaster	3/58
1061	6/24	41061	Midland Div	Derby	Manningham	6/55
1062	6/24	41062	Midland Div	Trafford Park	Bourneville	5/59
1063	7/24	41063	Midland Div	Trafford Park	Manningham	10/60
1064	7/24	41064	Midland Div	Trafford Park	Bourneville	1/57
1065	7/24	41065	Scotland	Kingmoor	Lancaster	3/56
1066	8/24	41066	Scotland	Kingmoor	Saltley	5/58
1067	8/24	41067	Scotland	Kingmoor	Manningham	2/55
1068	8/24	41068	Scotland	Kingmoor	Leeds	12/58
1069	9/24	41069	Scotland	Kingmoor	Gloucester	12/55
1070	9/24	41070	Midland Div	Leeds	Millhouses	12/55
1071	9/24	41071	Midland Div	Leeds	Leeds	3/58
1072	9/24	41072	Midland Div	Leeds	Millhouses	11/55
1073	10/24	41073	Midland Div	Durranhill	Bourneville	9/57
1074	10/24	41074	Midland Div	Kentish Town	Kentish Town	8/54
1075	10/24	41075	Midland Div	Durranhill	Manningham	4/57
1076	11/24	41076	Western Div	Millhouses	Rugby	5/55
1077	11/24	41077	Western Div	Millhouses	Rowsley	4/57
1078	11/24	41078	Western Div	Millhouses	Saltley	9/58
1079	11/24	41079	Western Div	Millhouses	Nottingham	11/56
1080	11/24	41080	Western Div	Corkerhill	Manningham	1/54
1081	12/24	41081	Western Div	Corkerhill	Lancaster	12/55

No.	Built	BR	First depot	1933 Allocation	Last depot	Withdrawn
1082	12/24	41082	Western Div	Corkerhill	Nottingham	4/54
1083	12/24	41083	Western Div	Polmadie	Derby	12/58
1084	12/24	41084	Western Div	Polmadie	Derby	6/54
1085	5/25	41085	Midland Div	Leeds	Bank Hall	1/57
1086	5/25	41086	Midland Div	Leeds	Derby	5/58
1087	6/25	41087	Midland Div	Leeds	Leeds	11/54
1088	6/25	41088	Midland Div	Leeds	Derby	12/56
1089	6/25	41089	Midland Div	Trafford Park	Bourneville	7/57
1090	7/25	41090	Midland Div	Trafford Park	Monument Lane	12/58
1091	7/25	41091	Midland Div	York	Bedford	4/55
1092	7/25	41092	Midland Div	Nottingham	Stranraer	8/53
1093	7/25	41093	Midland Div	Nottingham	Gloucester	7/58
1094	7/25	41094	Midland Div	Nottingham	Leeds	1/59
1095	8/25	41095	Midland Div	Nottingham	Gloucester	2/58
1096	8/25	41096	Midland Div	Nottingham	Nottingham	5/54
1097	8/25	41097	Midland Div	Nottingham	Nottingham	5/56
1098	8/25	41098	Midland Div	Derby	Leicester	6/57
1099	9/25	41099	Midland Div	Leicester	Stranraer	12/53
1100	9/25	41100	Midland Div	Leicester	Leeds	4/59
1101	9/25	41101	Midland Div	Leicester	Lancaster	8/59
1102	10/25	41102	Midland Div	Leicester	Blackpool	12/58
1103	10/25	41103	Midland Div	Leicester	Derby	12/57
1104	10/25	41104	Midland Div	Leicester	Leeds	8/55
1105	11/25	41105	Midland Div	Camden	Rugby	10/57
1106	11/25	41106	Midland Div	Camden	Chester	7/58
1107	11/25	41107	Midland Div	Camden	Lancaster	10/55
1108	11/25	41108	Midland Div	Camden	Lancaster	1/57
1109	12/25	41109	Midland Div	Camden	Dumfries	12/52
1110	12/25	41110	Western Div	Camden	Hurlford	8/54
1111	12/25	41111	Western Div	Chester	Holyhead	5/58
1112	12/25	41112	Western Div	Camden	Lancaster	11/57
1113	12/25	41113	Western Div	Bushbury	Lancaster	12/58
1114	12/25	41114	Western Div	Chester	Derby	5/58
1115	6/25	41115	Western Div	Crewe North	Holyhead	5/54
1116	7/25	41116	Western Div	Chester	Trafford Park	12/57
1117	8/25	41117	Western Div	Carnforth	Gloucester	5/55
1118	8/25	41118	Western Div	Crewe North	Trafford Park	1/58
1119	9/25	41119	Western Div	Crewe North	Leeds	12/58

No.	Built	BR	First depot	1933 Allocation	Last depot	Withdrawn
1120	10/25	41120	Western Div	Crewe North	Llandudno Jcn	6/59
1121	10/25	41121	Western Div	Crewe North	Derby	2/59
1122	11/25	41122	Western Div	Rugby	Rugby	12/58
1123	11/25	41123	Western Div	Carnforth	Gloucester	12/59
1124	11/25	41124	Western Div	Carnforth	Llandudno Jcn	1/55
1125	12/25	41125	Western Div	Crewe North	Perth	2/53
1126	12/25	41126	Western Div	Carnforth	St Rollox	11/56
1127	12/25	41127	Western Div	Carnforth	Stranraer	8/55
1128	12/25	41128	Western Div	Monument Lane	St Rollox	3/56
1129	12/25	41129	Western Div	Monument Lane	Stranraer	6/55
1130	1/26	41130	Western Div	Longsight	Carstairs	9/55
1131	1/26	41131	Western Div	Longsight	Stranraer	3/56
1132	2/26	41132	Western Div	Longsight	Stranraer	10/56
1133	2/26	41133	Western Div	Crewe North	Corkerhill	10/54
1134	2/26	41134	Western Div	Camden	Corkerhill	10/54
1135	6/25	41135	Scotland	Kingmoor	Stranraer	9/55
1136	6/25	41136	Scotland	Kingmoor	Lancaster	10/55
1137	7/25	41137	Scotland	Kingmoor	Leeds	4/56
1138	7/25	41138	Scotland	Kingmoor	Ayr	12/54
1139	7/25	41139	Scotland	Kingmoor	Corkerhill	10/54
1140	7/25	41140	Scotland	Kingmoor	Saltley	5/57
1141	7/25	41141	Scotland	Kingmoor	Kingmoor	9/54
1142	7/25	41142	Scotland	Kingmoor	Corkerhill	7/56
1143	7/25	41143	Scotland	Kingmoor	Derby	3/59
1144	7/25	41144	Scotland	Kingmoor	Nottingham	3/58
1145	7/25	41145	Scotland	Kingmoor	Carstairs	10/53
1146	7/25	41146	Scotland	Kingmoor	Kingmoor	11/54
1147	8/25	41147	Scotland	Kingmoor	Stranraer	3/56
1148	8/25	41148	Scotland	Kingmoor	Greenock	3/53
1149	8/25	41149	Scotland	Kingmoor	Greenock	8/55
1150	8/25	41150	Western Div	Chester	Trafford Park	10/57
1151	8/25	41151	Western Div	Chester	Lancaster	1/57
1152	9/25	41152	Western Div	Rugby	Lancaster	3/58
1153	9/25	41153	Western Div	Rugby	Bourneville	11/57
1154	9/25	41154	Western Div	Rugby	Trafford Park	8/55
1155	9/25	41155	Western Div	Rugby	Stranraer	3/57
1156	9/25	41156	Western Div	Rugby	Bourneville	9/58
1157	9/25	41157	Western Div	Edge Hill	Derby	5/60

No.	Built	BR	First depot	1933 Allocation	Last depot	Withdrawn
1158	9/25	41158	Western Div	Chester	Chester	9/59
1159	10/25	41159	Western Div	Edge Hill	Millhouses	4/58
1160	9/25	41160	Western Div	Monument Lane	Crewe North	10/56
1161	9/25	41161	Western Div	Bushbury	Trafford Park	12/55
1162	9/25	41162	Western Div	Bushbury	Rugby	6/60
1163	9/25	41163	Western Div	Bushbury	Manningham	12/58
1164	9/25	41164	Western Div	Bushbury	Trafford Park	10/58
1165	9/25	41165	Western Div	Bushbury	Bourneville	3/59
1166	9/25	41166	Western Div	Bushbury	Trafford Park	9/56
1167	10/25	41167	Western Div	Bushbury	Rugby	10/58
1168	10/25	41168	Western Div	Bushbury	Monument Lane	7/61
1169	10/25	41169	Western Div	Bushbury	Trafford Park	7/55
1170	11/25	41170	Western Div	Monument Lane	Trafford Park	9/56
1171	11/25	41171	Western Div	Carnforth	Dumfries	12/52
1172	11/25	41172	Western Div	Monument Lane	Rugby	7/57
1173	11/25	41173	Western Div	Monument Lane	Derby	2/59
1174	11/25	41174	Western Div	Monument Lane	Rugby	2/54
1175	11/25	41175	Western Div	Kingmoor	Dumfries	4/55
1176	11/25	41176	Western Div	Kingmoor	Corkerhill	1/55
1177	11/25	41177	Western Div	Kingmoor	Stranraer	12/55
1178	11/25	41178	Western Div	Kingmoor	Dalry Road	12/53
1179	12/25	41179	Western Div	Corkerhill	Dumfries	9/57
1180	12/25	41180	Western Div	Corkerhill	Saltley	3/57
1181	12/25	41181	Western Div	Ayr	Gloucester	11/57
1182	12/25	41182	Western Div	Ayr	Greenock	12/52
1183	12/25	41183	Western Div	Aberdeen	Ayr	2/55
1184	12/25	41184	Western Div	Aberdeen	Aberdeen	6/53
1185	2/27	41185	Central (L&Y)	Low Moor	Derby	11/57
1186	2/27	41186	Central (L&Y)	Low Moor	Southport	9/57
1187	2/27	41187	Central (L&Y)	Low Moor	Bank Hall	7/56
1188	3/27	41188	Central (L&Y)	Low Moor	Lower Darwen	12/55
1189	3/27	41189	Central (L&Y)	Southport	Lancaster	7/58
1190	3/27	41190	Central (L&Y)	Southport	Millhouses	1/58
1191	3/27	41191	Central (L&Y)	Bank Hall	Leeds	3/56
1192	3/27	41192	Central (L&Y)	Bank Hall	Kettering	6/57
1193	3/27	41193	Central (L&Y)	Bank Hall	Lancaster	11/58
1194	3/27	41194	Central (L&Y)	Blackpool	Bourneville	9/57
1195	3/27	41195	Central (L&Y)	Blackpool	Gloucester	12/57

No.	Built	BR	First depot	1933 Allocation	Last depot	Withdrawn
1196	3/27	41196	Central (L&Y)	Blackpool	Lancaster	7/58
1197	3/27	41197	Central (L&Y)	Blackpool	Lancaster	5/57
1198	3/27	41198	Central (L&Y)	Wakefield	Leicester	12/55
1199	4/27	41199	Central (L&Y)	Wakefield	Millhouses	1/58
900	4/27	40900	Scotland	Polmadie	Nottingham	4/56
901	4/27	40901	Scotland	Polmadie	Carstairs	6/54
902	4/27	40902	Scotland	Polmadie	Dumfries	8/56
903	4/27	40903	Scotland	Polmadie	Carstairs	9/55
904	4/27	40904	Scotland	Polmadie	Carstairs	3/57
905	4/27	40905	Scotland	Polmadie	Corkerhill	12/53
906	4/27	40906	Scotland	Polmadie	Corkerhill	1/55
907	4/27	40907	Scotland	Polmadie	Millhouses	10/60
908	5/27	40908	Scotland	Polmadie	Corkerhill	7/55
909	5/27	40909	Scotland	Polmadie	Corkerhill	7/56
910	6/27	40910	Scotland	Ayr	Trafford Park	6/56
911	6/27	40911	Scotland	Corkerhill	Dalry Road	12/52
912	6/27	40912	Scotland	Corkerhill	Dumfries	4/55
913	6/27	40913	Scotland	Corkerhill	Corkerhill	8/55
914	6/27	40914	Scotland	Corkerhill	Corkerhill	9/54
915	6/27	40915	Scotland	Hurlford	Hurlford	12/55
916	7/27	40916	Scotland	Hurlford	Polmadie	8/55
917	7/27	40917	Scotland	Carstairs	Bourneville	12/56
918	7/27	40918	Scotland	Aberdeen	St Rollox	12/52
919	7/27	40919	Scotland	Dalry Road	Corkerhill	1/54
920	7/27	40920	Scotland	Dalry Road	Stranraer	5/58
921	7/27	40921	Scotland	Perth	Perth	12/55
922	7/27	40922	Scotland	Perth	Perth	12/52
923	7/27	40923	Scotland	Perth	Perth	11/54
924	7/27	40924	Scotland	Perth	Perth	3/55
925	5/27	40925	Midland	Nottingham	Derby	12/59
926	5/27	40926	Midland	Nottingham	Crewe North	8/57
927	5/27	40927	Midland	Nottingham	Derby	7/57
928	7/27	40928	Midland	Nottingham	Saltley	3/58
929	5/27	40929	Midland	Durranhill	Rowsley	7/56
930	5/27	40930	Midland	Durranhill	Gloucester	4/57
931	5/27	40931	Midland	Durranhill	Rowsley	11/58
932	6/27	40932	Midland	Durranhill	Gloucester	5/56
933	6/27	40933	Midland	Derby	Monument Lane	4/58

No.	Built	BR	First depot	1933 Allocation	Last depot	Withdrawn
934	6/27	40934	Midland	Derby	Gloucester	3/57
935	9/32	40935	Derby	Derby	Bourneville	4/58
936	9/32	40936	Derby	Derby	Monument Lane	1/61
937	9/32	40937	Bank Hall	Bank Hall	Lancaster	4/58
938	9/32	40938	Aberdeen	Aberdeen	Perth	8/56
939	9/32	40939	Perth	Perth	Perth	11/56

LMS 2P 4-4-0

Dimensions

Cylinders	19in x 26in
Coupled wheel diameter	6ft 9in
Bogie wheel diameter	3ft 6½in
Boiler pressure	180lbs psi
Heating surface	1,410sq ft
Grate area	21.1sq ft
Axleload	17¾ tons
Weight Engine	54 tons 1 cwt
Tender	41 tons 4 cwt
Total	95 tons 5 cwt
Water capacity	3,500 gallons
Coal capacity	4 tons
Tractive effort	16,551lbs

Weight diagram

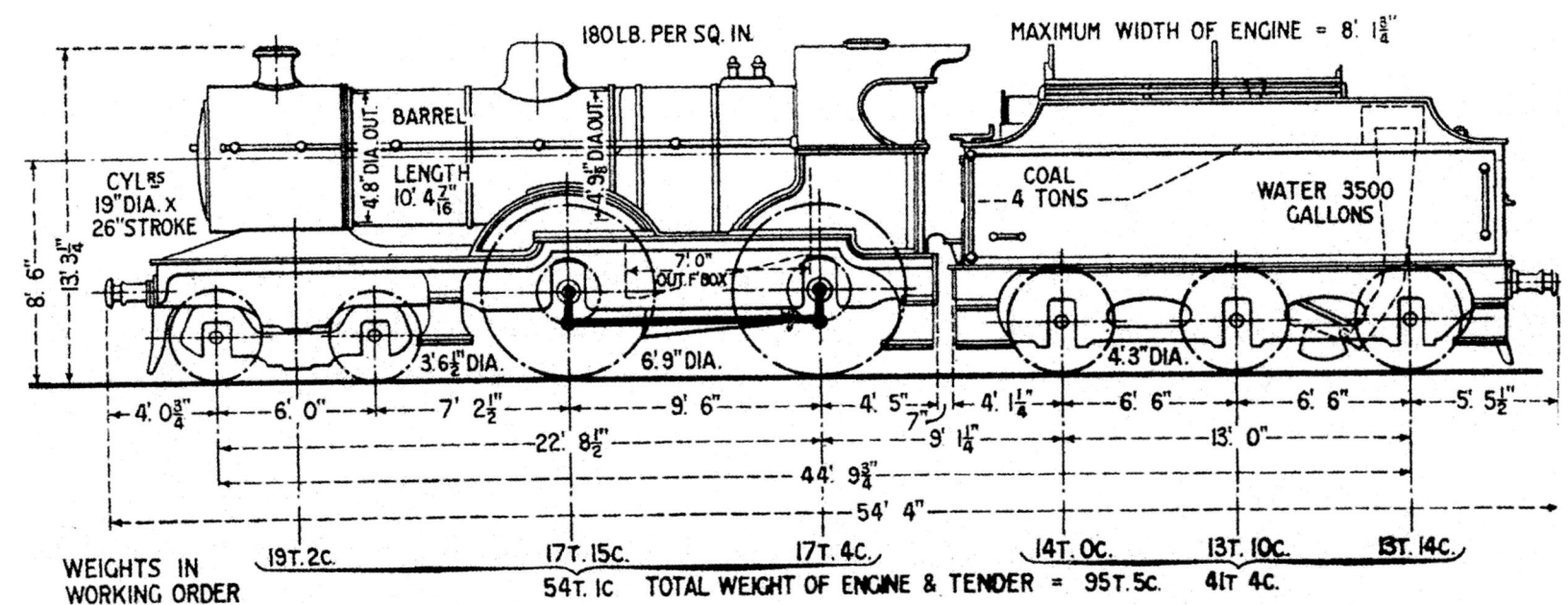

Statistics

No.	Built	BR	First depot ++	Last depot	Withdrawn	
563	1928	40563	Bath	Templecombe	5/62	
564	1928	40564	Bath	Templecombe	2/62	
565	1928	40565	Midland	Preston	11/59	
566	1928	40566	Midland	Stranraer	9/61	
567	1928	40567	Midland	Kentish Town	8/59	
568	1928	40568	Bath	Templecombe	2/59	
569	1928	40569	Bath	Templecombe	11/61	
570	1928	40570	Hurlford	Hurlford	8/61	
571	1928	40571	Hurlford	Hurlford	7/61	
572+	1928	40572	Hurlford	Hurlford	7/61	
573	1928	40573	Scotland	Hurlford	5/59	
574	1928	40574	Scotland	Hurlford	5/61	
575	1929*	40575	Scotland	Hurlford	8/61	
576	1929*	40576	Scotland	Dumfries	11/59	
577	1928	40577	Corkerhill	Dumfries	7/61	
578	1928	40578	Ardrossan	Ardrossan	12/61	
579	1928	40579	Ardrossan	Ardrossan	7/61	
580	1929*	40580	Central (L&Y)	Kentish Town	1961	
581	1928	40581	Central (L&Y)	Royston	9/60	
582	1928	40582	Central (L&Y)	Kentish Town	11/59	
583	1928	40583	Central (L&Y)	Stafford	9/60	
584	1928	40584	Central (L&Y)	Farnley Jcn	9/60	
585	1928	40585	Central (L&Y)	Nottingham	1961	
586	1928	40586	Central (L&Y)	Southport	4/61	
587	1928	40587	Central (L&Y)	Wigan	11/59	
588	1928	40588	Central (L&Y)	Bank Hall	11/60	
589	1928	40589	Central (L&Y)	Rhyl	11/59	
590	1928	40590	Scotland	Ayr	6/59	
591	1928	40591	Scotland	Hurlford	11/34	Accident damage
592	1928	40592	Scotland	Hurlford	12/61	
593	1928	40593	Scotland	Hurlford	8/61	
594	1928	40594	Scotland	Corkerhill	11/59	
595	1928	40595	Scotland	Hurlford	7/61	
596	1928	40596	Scotland	Corkerhill	9/61	
597	1928	40597	Scotland	Hurlford	8/61	
598	1928	40598	Scotland	Corkerhill	11/59	

No.	Built	BR	First depot ++	Last depot	Withdrawn	
599	1928	40599	Scotland	Corkerhill	5/59	
600	1928	40600	Midland	Keith	4/59	
601+	1928	40601	Bath	Bath	12/59	
602	1928	40602	Midland	Hurlford	10/61	
603	1928	40603	Scotland	Keith	7/61	
604	1928	40604	Scotland	Keith	7/61	
605	1928	40605	Scotland	Hurlford	10/59	
606	1928	40606	Ardrossan#	Ardrossan	5/59	
607	1928	40607	Scotland#	Dumfries	8/60	
608	1928	40608	Ardrossan#	Hurlford	9/59	
609	1928	40609	Ardrossan#	Hurlford	10/61	
610	1928	40610	Greenock	Ayr	5/59	
611	1928	40611	Greenock	Stranraer	9/59	
612	1928	40612	Scotland#	Hurlford	10/61	
613	1929	40613	Scotland	Corkerhill	10/61	
614	1929	40614	Scotland	Dumfries	10/61	
615	1929	40615	Scotland	Corkerhill	10/61	
616	1929	40616	Scotland	Stranraer	6/59	
617	1929	40617	Scotland	Keith	11/59	
618	1929	40618	Scotland	Keith	10/61	
619	1929	40619	Scotland	Aberdeen Ferryhill	10/61	
620	1929	40620	Greenock	Corkerhill	11/61	
621	1929	40621	Greenock	Corkerhill	10/61	
622	1929	40622	Scotland	Keith	6/61	
623	1929	40623	Scotland	Stranraer	8/61	
624	1929	40624	Ardrossan	Ardrossan	8/61	
625	1929	40625	Ardrossan	Ardossan	10/61	
626	1929	40626	Ardrossan	Hurlford	10/61	
627	1929	40627	Stirling	Corkerhill	3/61	
628	1929	40628	Midland	Carlisle Upperby	1/61	
629	1930	40629	Midland	Corkerhill	4/61	
630	1930	40630	Midland	Normanton	9/60	
631	1930	40631	Midland	Patricroft	9/60	
632	1930	40632	Midland	Nottingham	2/61	
633**	1928	40633	Bath	Burton	10/59	Dabeg feedwater heater
634**	1928	40634	Bath	Templecombe	5/62	

Dimensions, Weight Diagrams & Statistics • 339

No.	Built	BR	First depot ++	Last depot	Withdrawn	
635**	1928	40635	Bath	Llandudno Jcn	2/61	
636	1931	40636	Scotland	Corkerhill	10/59	
637	1931	40637	Scotland	Corkerhill	9/61	
638	1931	40638	Scotland	Ayr	6/62	
639	1931	40639	Scotland	Corkerhill	11/34	Accident damage
640	1931	40640	Scotland	Ayr	9/61	
641	1931	40641	Scotland	Hurlford	10/61	
642	1931	40642	Scotland	Hurlford	11/61	
643	1931	40643	Scotland	Hurlford	11/61	
644	1931	40644	Scotland	Hurlford	10/59	
645	1931	40645	Scotland	Hurlford	11/61	
646	1931	40646	Scotland	Bescot	5/62	
647	1931	40647	Ayr	Hurlford	11/61	
648	1931	40648	Scotland	Kittybrewster	10/61	
649	1931	40649	Scotland	Corkerhill	11/59	
650	1931	40650	Scotland	Kittybrewster	10/61	
651	1931	40651	Western	Kilmarnock	12/61	
652	1931	40652	Western	Templecombe	5/60	
653	1931	40653	Western	Crewe North	11/59	Dabeg feedwater heater
654	1931	40654	Western	Barrow	12/59	
655	1931	40655	Western	Crewe North	11/59	
656	1931	40656	Western	Carlisle Upperby	11/59	
657	1931	40657	Western	Watford	1962	
658	1931	40658	Western	Chester	11/59	
659	1931	40659	Western	Watford	9/61	
660	1931	40660	Western	Crewe North	11/59	
661	1931	40661	Scotland	Kilmarnock	12/61	
662	1931	40662	Scotland	Hurlford	9/54	
663	1931	40663	Scotland	Kittybrewster	10/61	
664	1931	40664	Scotland	Hurlford	5/62	
665	1931	40665	Scotland	Hurlford	6/62	
666	1932	40666	Stirling	Ardrossan	7/59	
667	1932	40667	Ardossan	Ardrossan	11/59	
668	1932	40668	Ardrossan	Ardrossan	10/61	
669	1932	40669	Polmadie	Ardrossan	9/61	
670	1932	40670	Polmadie	Ardrossan	11/62	
671	1932	40671	Western	Patricroft	11/60	

No.	Built	BR	First depot ++	Last depot	Withdrawn
672	1932	40672	Western	Watford	1962
673	1932	40673	Western	Bescot	11/59
674	1932	40674	Western	Longsight	11/59
675	1932	40675	Western	Stafford	8/59
676	1932	40676	Central (L&Y)	Patricroft	8/57
677	1932	40677	Central (L&Y)	Preston	11/59
678	1932	40678	Central (L&Y)	Bescot	10/61
679	1932	40679	Central (L&Y)	Crewe North	11/59
680	1932	40680	Central (L&Y)	Wigan	11/59
681	1932	40681	Central (L&Y)	Patricroft	5/62
682	1932	40682	Central (L&Y)	Nottingham	10/61
683	1932	40683	Central (L&Y)	Watford	10/61
684	1932	40684	Central (L&Y)	Bank Hall	10/61
685	1932	40685	Central (L&Y)	Hellifield	7/61
686	1932	40686	Scotland	Hurlford	10/61
687	1932	40687	Scotland	Hurlford	11/61
688	1932	40688	Scotland	Hurlford	1/60
689	1932	40689	Scotland	Hurlford	11/61
690	1932	40690	Central (L&Y)	Leeds	9/60
691	1932	40691	Central (L&Y)	Nottingham	2/61
692	1932	40692	Western	Bescot	8/61
693	1932	40693	Western	Longsight	6/59
694	1932	40694	Western	Bescot	5/62
695	1932	40695	Western	Hurlford	3/61
696	1932	40696	Bath	Bath	5/62
697	1932	40697	Bath	Bath	2/62
698	1932	40698	Bath	Bath	7/60
699	1932	40699	Bath	Llandudno Jcn	4/60
700	1932	40700	Bath	Bath	5/62

* Replacements for originals sent to Somerset & Dorset Railway
** From S&D Railway in 1930
\+ 572 fitted with double port exhaust valves and renumbered 601
\# Initially allocated to Midland Division but transferred to Scotland shortly afterwards
++ where initial depot allocations unknown, the locos were allocated to the following depots:
   Midland Division: Bath, Templecombe, Leeds Holbeck, Sheffield Millhouses
   Western Division: Workington, Carlisle Upperby, Rhyl, Chester, Crewe North, Northampton
   Central Division: Bank Hall, Southport, Newton Heath, Bury, Low Moor, Wakefield
   Northern Division (Scotland): Polmadie, Corkerhill, Hurlford, Ardrossan, Ayr, Stranraer, Girvan, Greenock

Locomotives proposed but not built
Deeley 4-cylinder 4-6-0 Compound (1907)
Dimensions

High pressure cylinders (2)	13in x 28in
Low pressure cylinders (2)	21in x 26in
Coupled wheel diameter	6ft 6½in
Inside Walschaerts valve gear	
Boiler pressure	220lbs psi
Tractive effort	27,400lbs

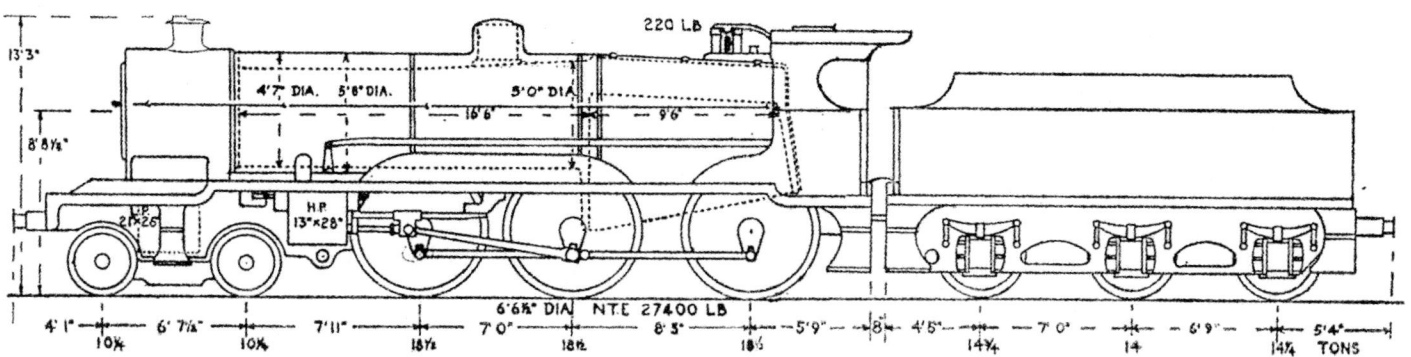

Deeley proposal (1907) for 4-cylinder compound 4—6—0

A.R.L.E. Maunsell/Fowler 4-4-0 (1918)
Dimensions

Cylinders (2)	21in x 28in
Coupled wheel diameter	6ft 8in
Bogie wheel diameter	3ft 6in
Boiler pressure	200lbs psi
Grate area	28.4sq ft
Tractive effort	24,900lbs

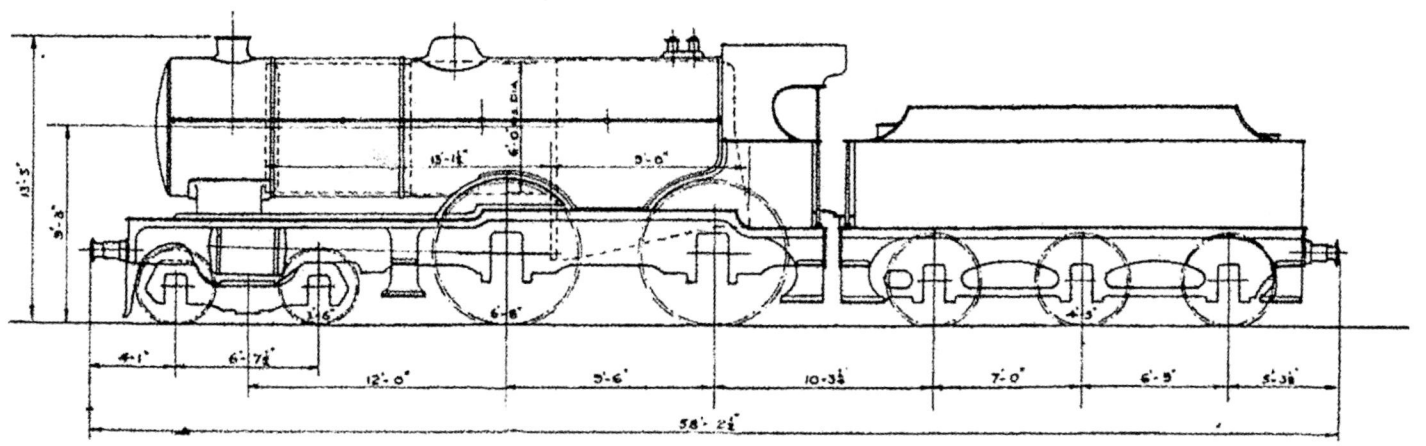

A.R.L.E. Maunsell/Fowler 4-6-0 (1918)

Cylinders (2)	23½in x 30in
Coupled wheel diameter	6ft 8in
Bogie wheel diameter	3ft 6in
Boiler pressure	220lbs psi
Grate area	31.5sq ft
Tractive effort	33,000lbs

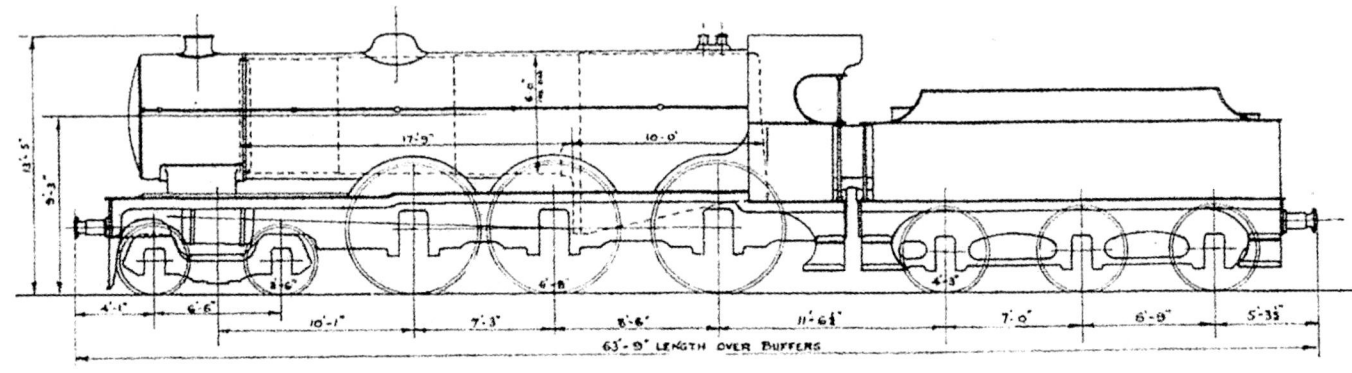

Derby 3-cylinder 4-6-0 Compound (1924)

Dimensions

High pressure cylinder (1)	20¼in x 30in
Low pressure cylinders (2)	22¼in x 30in
Coupled wheel diameter	6ft 9in
Bogie wheel diameter	3ft 3½in
Walschaerts valve gear	
Boiler pressure	220lbs psi
Heating surface	2,494sq ft (incl 529sq ft superheater)
Grate area	31.5sq ft
Axleload	20 tons
Total Weight	124 tons 14 cwt
Water capacity	3,500 gallons
Coal capacity	5½ tons
Tractive effort	32,270lbs

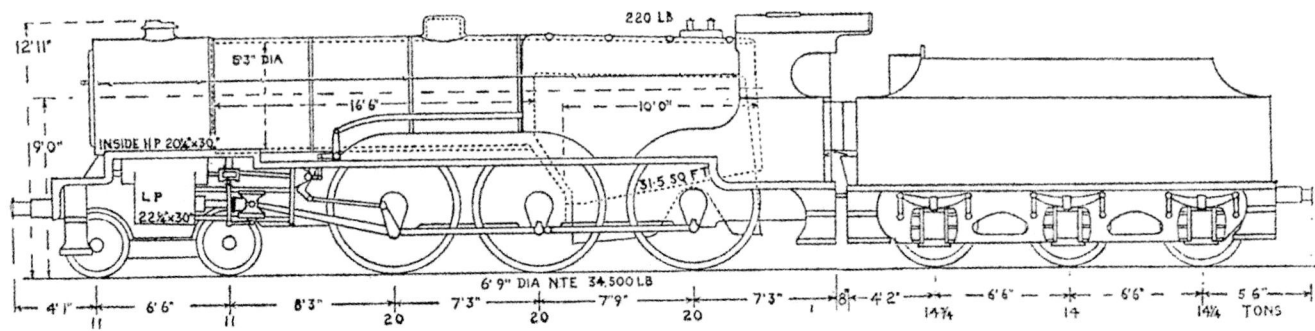

Derby proposal (1924) for 3-cylinder compound 4—6—0

Derby 4-cylinder 4-6-2 Compound (1926)

Dimensions

High pressure cylinder (2)	16¾in x 26in
Low pressure cylinders (2)	23¾in x 26in
Coupled wheel diameter	6ft 9in
Bogie wheel diameter	3ft 3½in
Walschaerts valve gear	
Boiler pressure	240lbs psi
Heating surface	3,209sq ft (incl 631sq ft superheater)
Grate area	43.5sq ft
Axleload	21 tons
Total Weight	143 tons 14 cwt
Water capacity	3,500 gallons
Coal capacity	5½ tons
Tractive effort	34,600lbs

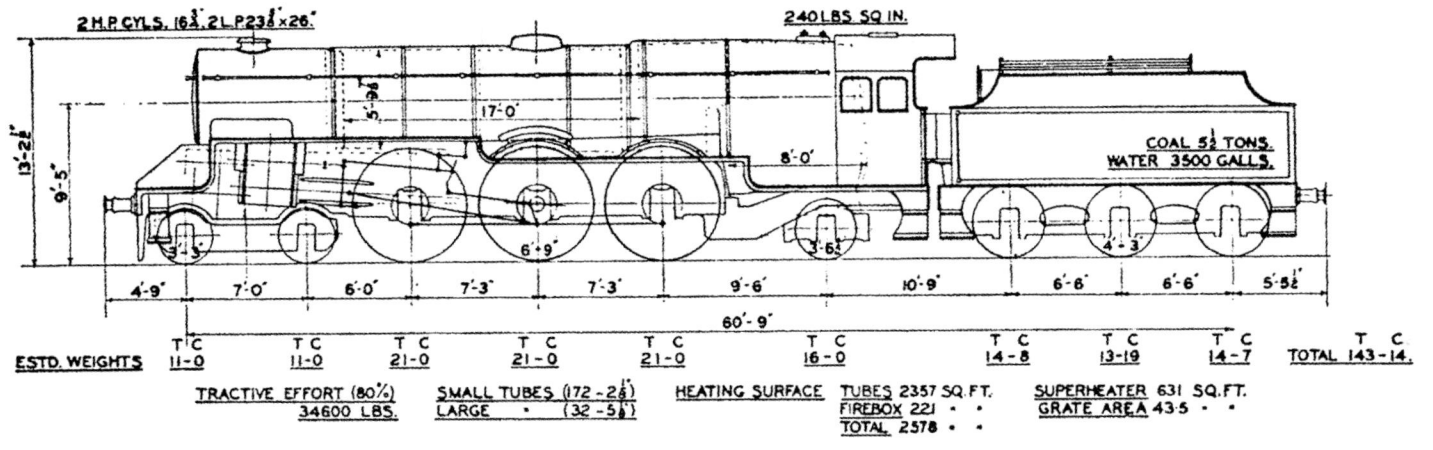

MIDLAND MAIN LINE GRADIENT CHARTS

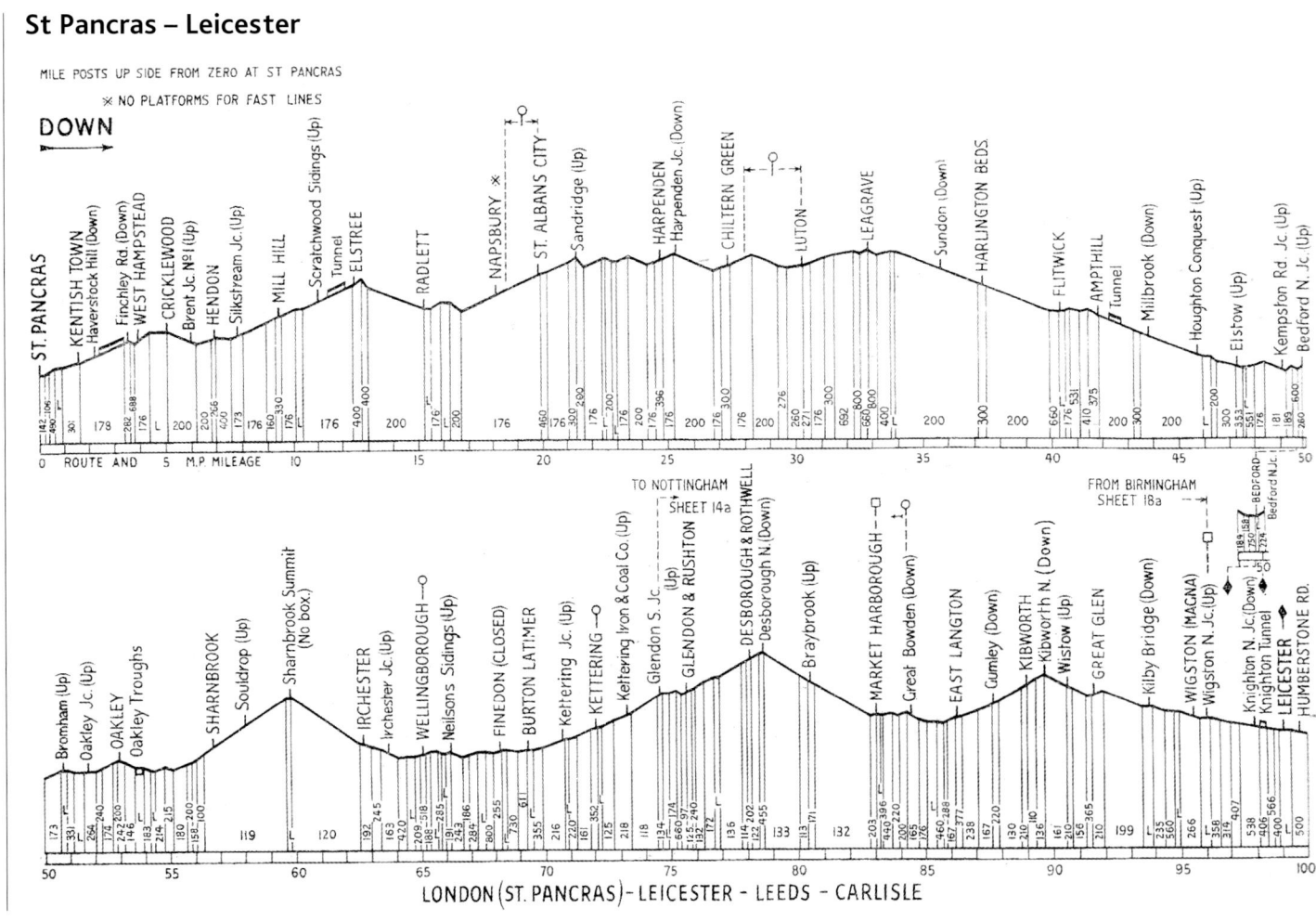

St Pancras – Leicester

Leicester – Leeds

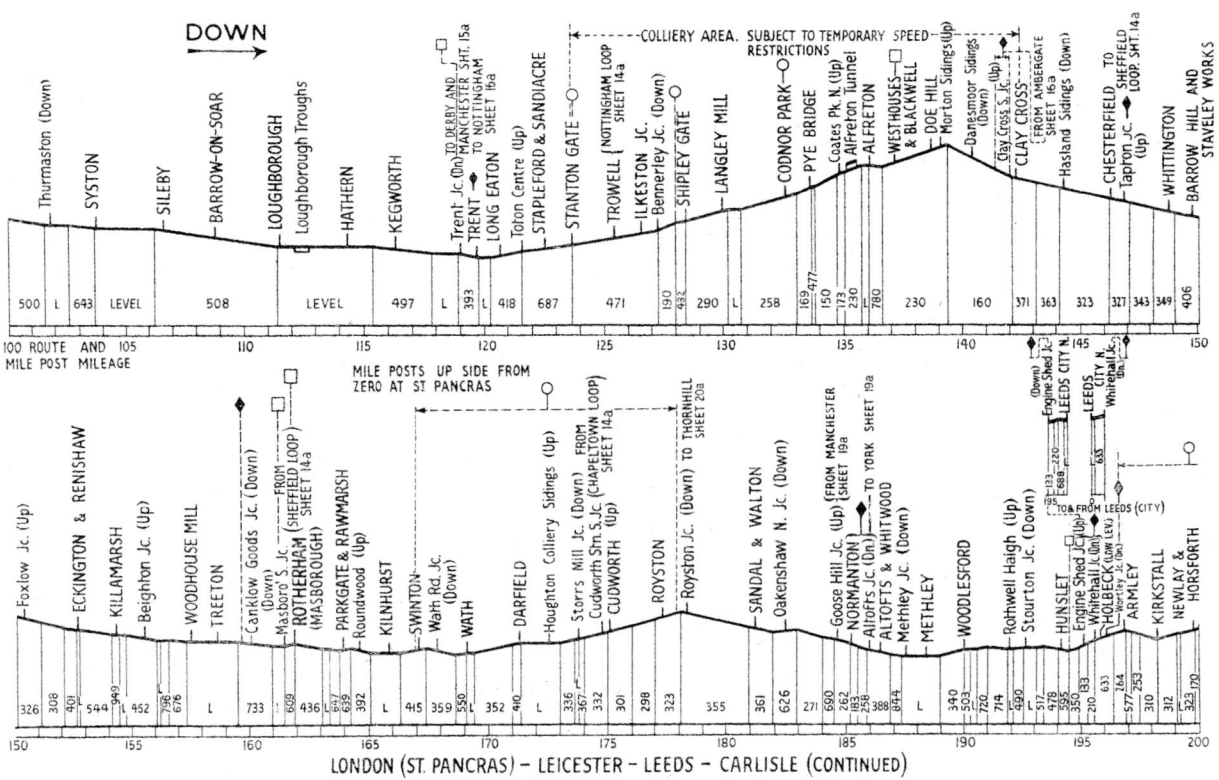

Leeds – Blea Moor

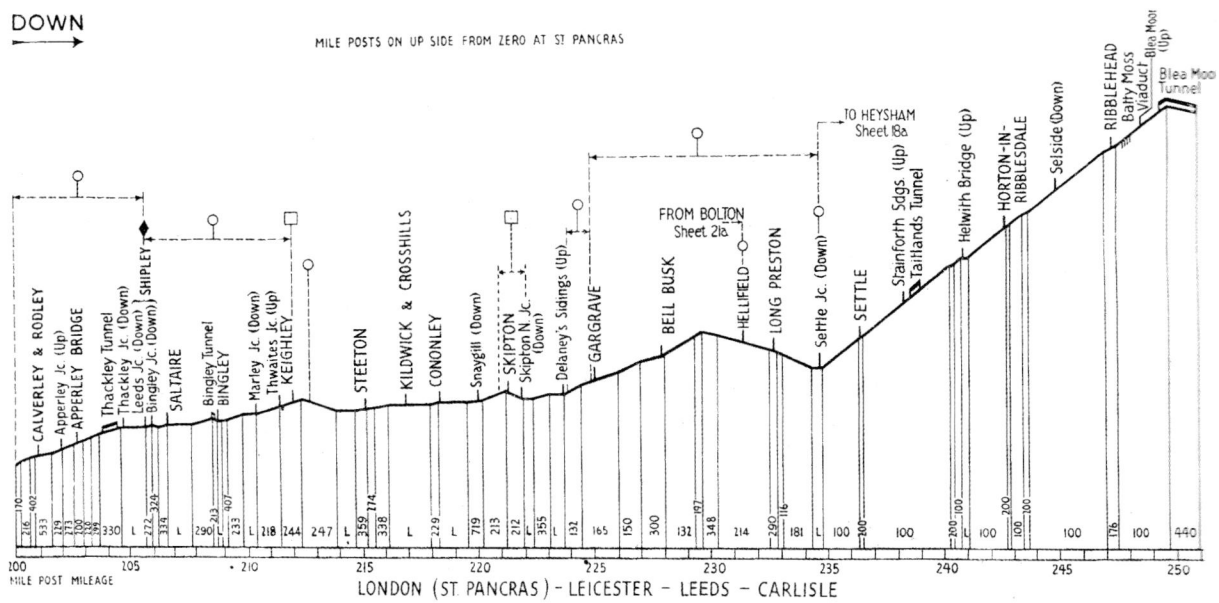

Blea Moor - Carlisle

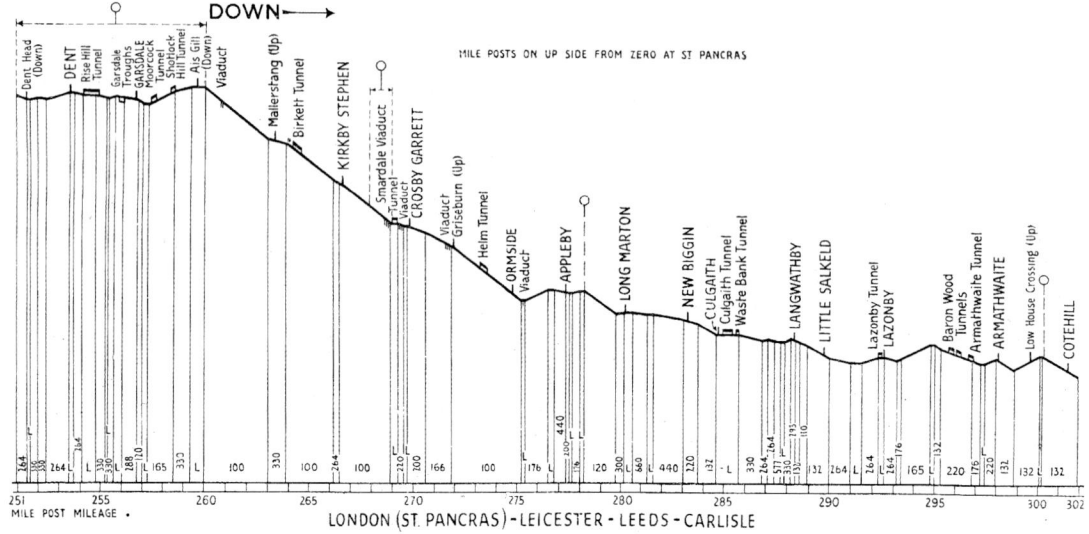

Derby - Manchester

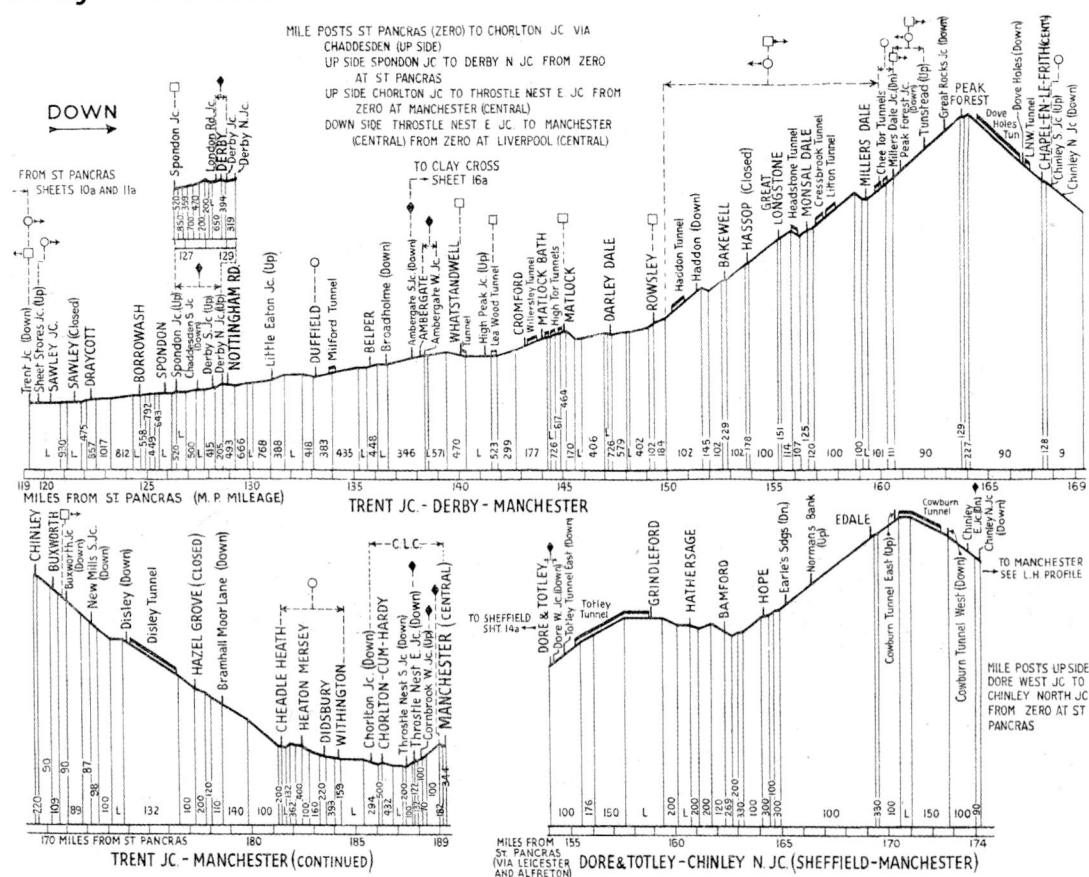

BIBLIOGRAPHY

Aves, W.A.T., *The LMS Compounds in Scotland,* Steam Days, Redgauntlet Publications, March 1998.
Aves, W.A.T., *The LMS '2P' 4-4-0s 1928 to 1962,* Steam Days, Redgauntlet Publications, October 2003.
Becket, W.S., *The Xpress Locomotive Register, Vol 2 London Midland Region,* Xpress Publishing, 1998.
Binns, Donald, *Midland and LMS Compounds,* Trackside Publications, 1996.
Bradley, D., and Milton, David, *Somerset & Dorset Locomotive History,* David & Charles, 1973.
Braithwaite, Jack, *S.W. Johnson Midland Railway Locomotive Engineer Artist,* Wyvern Publications, 1985.
Essery, R.J. and Jenkinson, D., *An Illustrated Review of Midland Locomotives, Volume 2,* Wild Swan, 1988.
Essery, R.J. and Jenkinson, D., *An Illustrated Review of LMS Locomotives, Volume 4,* Silver Link, 1987.
Haresnape, Brian, *Fowler Locomotives, A Pictorial History,* Ian Allan, 1972.
Higgs, Tony, *The Monument Lane 4-4-0 Compounds,* Steam Days, Redgauntlet Publications, April 2019.
Longworth Hugh, *BR Steam Locomotives complete allocations history 1948-1968,* Oxford Publishing Co., 2014.
Nock, O.S., *LMS Steam,* David & Charles, 1971.
Nock, O.S., *The Midland Compounds,* David & Charles Locomotive Monographs, 1964.
RCTS, *Locomotives of the LNER Part 4 (D52-D54),* RCTS 1968.
Summerson, Stephen, *Midland Railway Locomotives, Volume 1, General Survey,* Irwell Press, 2000.
Summerson, Stephen, *Midland Railway Locomotives, Volume 3, Johnson Classes,* Irwell Press.
Summerson, Stephen, *Midland Railway Locomotives, Volume 4, Deeley & Fowler Classes,* Irwell Press.
Tuplin, W.A., *Midland Steam,* David & Charles, 1973.

INDEX

Association of Railway Locomotive Engineers (ARLE), 209

Bibliography, 347

Engineers & Officers
Anderson, J., 209-211, 222
Churchward, G.J., 198, 204, 209, 266, 299, 301
Deeley, R.M, 12-13
Fowler, Sir Henry, 13-14
Hughes, G, 204, 210-211
Johnson, S.W, 12
Maunsell, R.E., 209, 265, 301
Paget, Cecil, 171, 209, 300
Schmidt (superheaters), 196
Smith, W.M, 158, 163, 299
Stanier, Sir William, 266, 301

Gradient charts
Blea Moor – Carlisle, 346
Derby – Manchester, 346
Leeds – Blea Moor, 345
Leicester – Leeds, 345
St Pancras – Leicester, 344

Locomotives (by class)
Midland 2P
1312 (300-309)
Dimensions & construction, 15, 302
Operation, 18
Statistics, 303
Weight diagram, 302
Withdrawal, 15

1327 (310-327)
Dimensions & construction, 20, 302
Operation, 22
Statistics, 303-304
Weight diagram, 303
Withdrawal, 20, 25

1562 & 1667 (328-357)
Dimensions & construction, 25-26, 29, 304
Operation, 31-32, 34, 38
Statistics, 305-306
Weight diagram, 304
Withdrawal, 31, 38

1738 (358-377)
Dimensions & construction, 38, 40, 306
Operation, 42-43, 45
Statistics, 307-308
Weight diagram, 307
Withdrawal, 40, 45

1808 (378-402)
Dimensions & construction, 45, 48, 308
Operation, 50
Statistics, 308-309
Withdrawal, 48

2183 (403-427)
Dimensions & construction, 57, 61, 309
Operation, 61, 63-64, 66
Statistics, 310-311
Weight diagram, 310
Withdrawal, 61

2203 (428-472)
Dimensions & construction, 66, 71, 311
Operation, 71-72, 78
Statistics, 312-313
Weight diagram, 311
Withdrawal, 71, 79

2581 (473-482)
Dimensions & construction, 79, 81, 313
Operation, 81
Statistics, 313
Withdrawal, 81

1667 (replacement) (483-492)
Dimensions & construction, 82, 314
Operation, 86
Statistics, 315
Weight diagram, 314
Withdrawal, 86

150 (493-502)
Dimensions & construction, 89, 91, 315
Operation, 91
Statistics, 315
Withdrawal, 91, 93

2421 (503-522)
Dimensions & construction, 93, 316
Operation, 94, 97
Statistics, 316
Withdrawal, 94, 97

60 (523-562)
Dimensions & construction, 97-98, 317
Operation, 102-104, 107
Statistics, 317-318
Withdrawal, 98, 106

3P 'Belpaire' (700-779)
Dabeg water heater, 129, 132, 136
Dimensions & construction, 124, 129, 322-323
Livery, 129
Oil burning, 129-130
Operation, 136, 138-139, 145, 150, 155
Performance, 138-139, 142
Statistics, 324-326
Tests, 142, 196
Weight diagram, 323
Withdrawal, 129, 136, 155

4P Midland 'Compound' (1000-1044)
Dimensions & construction, 158, 161, 163, 165, 326
Livery, 165
Maintenance, 165
Oil burning, 165
Operation, 168-169, 178, 184, 188
Performance, 169, 172, 178, 181
Preservation, 191-192, 194
Statistics, 328-329
Tests, 158, 160, 168, 170-171, 178, 181
Weight diagram, 327
Withdrawal, 165, 168, 1885

4P 'Simple' (990-999)
Accidents, 200
Dimensions & construction, 194, 196, 329
Operation, 198
Performance, 200
Statistics, 330
Tests, 196, 198, 201-202
Weight diagram, 329
Withdrawal, 198

M&GN (D52-D54)
Dimensions & construction, 120-121, 320-321
Statistics, 321-322
Withdrawal, 121

S&D 2P
Dimensions & construction, 109, 113, 116, 319
Operation, 120
Statistics, 320
Withdrawal, 120

LMS 4P 'Compound (1045-1199, 900-939)
Dimensions & construction, 211-212, 330
Enlarged tenders, 213, 216, 227-228, 232, 249, 255
Euston – Edinburgh non-stop, 213, 227
Livery, 213
Oil burning, 212-213
Operation, 222, 224, 230, 236-237, 246-248
Performance, 222, 227, 229, 237, 253-254
Statistics, 331-336
Tests, 201-202, 210, 222-223
Weight diagram, 330
Withdrawal, 213, 254

LMS 2P (563-700)
Accident, 269
Dabeg water heater, 268, 271, 273
Dimensions & construction, 265-266, 336
Livery, 269
Operation, 276-277
Performance, 276-277, 281-282, 292, 295
Statistics, 337-340
Weight diagram, 336
Withdrawal, 269, 296-29715

Locomotive proposals (not built)
ARLE Maunsell/Fowler 4-4-0, 341
ARLE Maunsell/Fowler 4-6-0, 342
Deeley 4-cyl compound 4-6-0, 341
Fowler 3-cyl compound 4-6-0, 342
Fowler 4-cyl compound 4-6-2, 343

Locomotives (Other)
Caledonian Dunalastairs, 206
GSWR Glenarthur 4-4-0, 206, 209
GW Castles, 13-14, 210, 212, 224
GW Saints & Stars, 204
Highland Castles, 206
LMS Royal Scots, 14, 210
LNWR Claughtons, 204-205
LNWR Prince of Wales, 204
L&Y Hughes 4-6-0s, 204-205
Midland Kirtley 2-4-0, 217

Logs
Bath – Evercreech Junction, 276, 283
Bedford – St.Pancras, 252-253Carlisle – Leeds, 140-141, 192, 202
Birmingham – Euston, 182
Carlisle – Carstairs - Glasgow, 224-225, 227
Cheltenham – Birmingham, 75, 96, 144-145, 238
Derby – Manchester, 33, 63, 142, 176
Euston – Birmingham, 229, 281
Forfar – Perth, 236-237
Glasgow St Enoch – Carlisle, 225-226
Hellifield – Carnforth, 194
Leeds – Carlisle, 140, 199, 201
Leeds – Crewe, 254
Leeds – St.Pancras, 141, 173
Leicester – St.Pancras, 33-34, 43, 102, 106, 143-144, 172, 175, 239
Manchester – Derby, 143, 176-177
Manchester – Liverpool, 155, 279
Nottingham – St.Pancras, 174
Preston – Carlisle, 223
Shrewsbury – Crewe, 238
St.Pancras – Leeds, 198-199
St.Pancras – Leicester, 42, 103, 138-139, 170-171

Personal experiences, 8-10, 253

Photographs – Locations
Accrington, 219
Aintree, 149, 262
Altrincham, 261
Ambergate, 153
Annan, 233
Armathwaite, 200
Attenborough, 59
Ayr, 256
Baguley, 79
Barnt Green, 95
Barton & Broughton, 156
Bath Green Park, 108, 113-115, 117, 268
Beattock, 231
Bedford, 57, 68, 105, 166-168
Binegar, 286
Birmingham New St., 31
Blaby, 86
Blackpool, 244, 264
Blackwell, 252
Bolton, 220, 297
Bourne End, 245
Bournemouth West, 8, 109, 111, 278
Bradford Manningham, 28
Bristol, 82, 98, 134
Burton, 54
Bushey, 228, 231, 234
Buxton, 69, 119
Buxworth Junction, 35, 91, 257

Camden, 205, 215
Carlisle, 74, 76, 126, 200, 203, 242, 261, 272, 275, 293
Carpenders Park, 228
Carstairs, 221
Castleton, 297
Chapel-en-le-Frith, 92, 148, 186, 240, 244
Chatburn, 288
Cheadle Heath, 95
Cheltenham, 107, 137, 185
Chester, 219
Chilcompton, 99, 110, 283
Chinley, 23, 37, 56, 65, 76-77, 88, 153-154, 157, 167, 190, 245, 249, 260, 262, 282
Clapham Junction (Yorks), 250
Colwyn Bay, 92
Crewe, Front cover, 70, 108, 249, 271
Cromer, 120
Cromford, 232

Darley Dale, 242
Delamere, 290
Derby, 11, 16, 24, 29-30, 41, 48-49, 53, 58, 60, 69-71, 101, 112, 119, 129, 132, 135, 147, 156, 165, 185, 196, 197, 216-218, 221, 267
Didsbury, 260
Diggle, 256, 264
Dillicar, 280
Doncaster, 193
Dore & Totley, 73
Dove Holes, 184
Duffield, 183
Dumfroes, 295

Elstree, 104, 145-147, 178, 235, 273

Farnworth, 56
Farrington Grove, 152

Garsdale, 44
Glasgow St.Enoch, 220, 226, 270, 275, 294
Glazebrook, 289
Gloucester, 60, 89
Goostrey, 258
Gow Hole, 189
Great Rocks, 94, 187

Hartford, 188
Headstone Tunnel, 232
Hellifield, 132
Hendon, 61
Highbridge, 112
Hough Green, 19
Hunts Cross, 51

Inverness, 207, 251

Kenilworth, 245
Kentish Town, 21, 41, 46, 80, 100, 125, 127, 161, 164, 239
Kilmarnock, Front cover jacket inset, 243, 246, 293, 295,
Kings Norton, 20, 55

Lazonby, 259
Leamington Spa, 218
Leeds, 17, 67, 81, 128, 131, 135, 162, 181, 217, 245
Leicester, 23, 163, 195
Lickey, 35, 75, 233, 257
Liverpool Brunswick, 16, 74, 90
Liverpool Lime St., 290
Llandudno Junction, 279
Long Eaton, 36
Lune River, 22
Luton, 190

Manchester Belle Vue, 47
Manchester Central, 59, 62, 85, 118, 157, 180
Manchester London Road, 37, 241, 292
Manchester Longsight, 214, 292
Manchester Trafford Park, 89, 193
Manchester Victoria, 51-52, 263
Marple, 259

Masbury, 285-286, 298
Mersey River, 289
Middleton Junction, 235
Milford Tunnel, 151
Mill Hill, 43, 62, 82, 136-137, 180

New Mills, 36, 77-78, 189, 240, 248
Newton Heath, 46
Northfield, 63
Nottingham, 85, 122, 132, 163

Peak Forest, 155
Perth, 208, 215
Poole, 284
Poulton-le-Fylde, 151
Prestatyn, 280

Radstock, 110, 285
Ratcliffe Junction, 87, 150
Rhyl, 90
Romiley, 187, 241
Rowsley, 278
Rugby, 25, 234, 274

St.Albans, 186
St.Pancras, 32, 99, 148, 152, 183
Stafford, 255
Saltley, 272
Sheffield, 30, 72, 118, 134
Shipley, 258
Shotlock Tunnel, 24
Skipton, 73, 101
Strines, 250-251
Tebay, 87
Templecombe, 284
Thurmaston, 122, 179
Timperley, 44
Totley Tunnel, 53
Toton, 21

Whitmore, 246
Wickwar, 55
Wigan, 49
Wigston, 179
Willesden Junction, 9

Yarmouth Beach, 123
York, 52, 68, 80

Photographs – Locomotives (by class)
1312 (300-309)
300, 17
301, 18
306, 17, 19
308, 19
1312, 16
1314, 15
1320, 16
1327 (310-327)
311, 25
322, 24
323, 20, 21, 118
325, 23
327, 24
1332, 21
1335, 22
1338, 23
40323, 119
40326, 119

1562/1667 (328-357)
331, 30
332, 37, 115
335, 29
338, 36
342, 30
343, 35
354, 35
1562, 26
1579, 27
1666, 26, Back cover
1667, 32
1668, 28
1670, 27
1673, 28
40332, 31
40337, 37

1738 (358-377)
360, 44
361, 41
368, 41
373, 40
1743, 38
1752, 39, 43
1757, 39

40377, 44

1808 (378-402)
14, 46
378, 48
381, 50-51
382, 53
383, 50, 52
384, 52
385, 56
386, 51
391, 53
392, 47
395, 55
397, 54-55
400, 47
1814, 45
1819, 46
40383, 49
40396, 56
40397, 49

(2183 (403-427)
403, 59, 64
406, 59
409, 65
414, 62
426, 62
2184, 62
2193, 57
2195, 61
2201, 58
2202, 57-58
40411, 65-66
40412, 11
40418, 60
40426, 60

2203 (428-472)
233, 67
235, 67
238, 73
428, 68
430, 74
431, 75
432, 74
433, 74
M 436, 70

437, 69
438, 78
446, 76
447, 69
461, 77
462, 77
463, 76
468, 282
2206, 73
2209, 68, 72
40443, 70
40454, 71
40464, 79

2581 (473-482)
473, 82
476, 81
2587, 80
2590, 82

1667 (replacement) (483-492)
483, 83-84
488, 85, 87
490, 85
491, 87
40487, 88
40489, 89
40491, 85, 88

150 (493-502)
209, 89
494, 82
495, 91
498, 90
499, 92
40495, 90

2421 (503-522)
503, 95
508, 93, 96
509, 97
521, 95
2424, 94
2425, 94
40518, 93

60 (523-562)
63, 99
65, 99
169, 98
523, 55, 105
550, 100, 104
554, 105
559, 100
40523, 107
40528, 108
40537, 108
40538, 101
40562, 101

M& GN (D52-D54)
1, 121
3, 121
14, 120
052, 123
53, 122
54, 123
77, 122

S&D 2P
18, 109, 112
41, 117
45, 110-111
67, 111
68, 110
69, 116
71, 114, 117
77, 114
301, 112
303, 113
321, 115

3P 'Belpaire' (700-779)
700, 131
702, 146
706, 132
708, 151
710, 149
711, 154
715, 128, 148
725, 53
726, 133
729, 134
730, 127
738, 148
740, 128
745, 153
746, 152
748, 154
749, 147
753, 146
755, 152
758, 127
759, 132, 153, 156
763, 147
767, 145
768, 130
773, 133, 150
774, 155
776, 129, 151
777, 130, 148, Back cover
779, 133
810, 126
859, 137
863, 126
2606, 136
2607, 125
2788, 125
2789, 137
40726, 157
40728, 135, 156
40741, 134
40758, 135

4P Midland 'Compound' (1000-1044)
1000, 164, 166, 193
1003, 162, 188
1005, 165
1007, 166
1009, 167, 179, 187
1012, 179, 183
1013, 161, 178
1014, 162
1017, 163, 180
1020, 187
1021, 65
1023, 181
1024, 186
1027, 183
1028, 185
1029, 185
1034, 186
1040, 164
1044, 184
2631, 160, 180
2632, 159
2633, 169
2634, 160
41003, 189
41007, 190
41009, 168
41014, 191
41016, 167, 190

4P 'simple' (990-999)
803, 197
807, 197
992, 196, 200, 203
995, 195
999, 195

LMS 4P 'Compound' (1045-1199, 900-939)
914, 233
923, 215
928, 240
M935, 218
936, 216, 249
1045, 183, 212
1046, 240
1048, 250
1050, 232
1052, 231
1053, 239
1054, 228
1057, 232
1061, 216
1066, 248
1069, 226
1072, 217, 244
1086, 217
1087, 242
1088, 245
1092, 245-246
1093, 241
1095, 242
1098, 233
1111, 214
1112, 213
1113, 246
1121, 241

1135, 231
1136, 243
1137, 214
1144, 249-250
1150, 234
1152, 215
1157, 218, 234, 243
M1162, 219
1166, 236
1176, 231
1185, 235
1189, 235
1195, 244
40903, 221
40906, 256
40913, 251
40920, 220
40936, 255
41050, 262
41052, 257
41063, 264
41066, 262-263
41074, 252
41076, 251
41077, 261
41085, 220
41097, 257
41100, 258, 264
41101, 264
41102, 263
41113, 9, 258
41119, Front cover, 261
41123, 260
41140, 259
41154, 259
41159, 260
41162, 10
41185, 256
41192, 221
41194, 219

LMS 2P (563-700)
563, 266
572, 267
574, 267
576, 268
591, 270
633, 268
634, 283
635, 269, 278
651, 279
653, 271
658, 280
663, 270
665, 277
677, 288
681, 288
693, 282
695, 280
697, 271
698, 278
700, 272
40563, 285-286
40564, 298
40565, 261
40568, 273, 286
40570, 293
40574, 295
40578, 275
40588, 297
40601, 8, 285
40609, 294
40613, 272, 275, 293
40616, 29540633, 273
40634, 284, Front cover inside jacket
40658, 290
40661, Front cover inside jacket
40667, 294
40674, 292
40677, 274

40679, 289
40683, 289
40684, 298
40685, 287
40693, 290, 292
40696, 284

Locomotives (Other)
Caledonian
63 (14630), 207

G&SWR
394, 209
14674, 208

GWR
Castles, 210, 224

Highland
55 (Clan), 208
140 (14765), 207

L&NWR
208 (6024), 205

L&Y
10445, 205

Midland
Johnson Singles, 86, 299-300

S&D
5 (1504), 112

LMS
5276, 154
5279, 244
5283, 96
5514, 234
6138, 245

6157, 64
BR (LMR)
43896, 259
43963, 260
44839, 97
45006, 262
45169, 295
45187, 10
45534, 298
45597, 66
45626, 89
45682, 156
46240, 293
53810, 298

BR (SR)
34099, 286
BR (ER)
60940, 293
61265, 262

BR
73049, 108

Train Timers
Allen, Cecil J, 42, 103, 142, 224, 300
Barrie, D.S, 239, 276, 281-282, 291
Bland, M.N, 248
Charlewood, R.E, 33-34, 103, 140, 299
Nathan, B.I, 254
Nock, O.S, 236
Rail Performance Society, 7, 239, 281, 291, 296
Rous-Marten, 138-139, 169, 299
Twibell, D, 192, 252-253